BASIC MATHEMATICS REVIEW

Loyce C. Gossage
Faculty Emeritus
Mt. San Antonio College
Walnut, California

SOUTH-WESTERN PUBLISHING CO.

Copyright © 1994
by SOUTH-WESTERN PUBLISHING CO.
Cincinnati, Ohio
ALL RIGHTS RESERVED

The text of this publication, or any part thereof, may not be reproduced or transmitted in any form or by any means, electronic or mechanical, including photocopying, recording, storage in an information retrieval system, or otherwise, without the prior written permission of the publisher.

ISBN: 0-538-70681-3
1 2 3 4 5 DH 97 96 95 94 93
Printed in the United States of America

Editor-in-Chief: Peter McBride
Managing Editor: Eve Lewis
Developmental Editor: Judith A. Witt
Coordinating Editor: Patricia Matthews Boies
Marketing Manager: Carol Ann Dana
Designer: Barbara Libby
Production Editor: Gayle J. Statman

To the Instructor

Why should a textbook in basic mathematics be revised every few years? Because mathematics, even basic mathematics, is not stagnant; it is continually changing and expanding. Admittedly, basic mathematics does not expand as rapidly as higher mathematics, but it must change because wages, prices, taxes, interest rates, etc., do change, and different concepts move to the forefront.

In addition to containing the usual changes to keep the text realistically compatible with current and future economic problems pertaining to wages, prices, and the like, this edition of *BASIC MATHEMATICS REVIEW* contains some new lessons. These are designed to introduce simple estimation and to help students better understand decimals and percents. Also new in this edition are additional pages of review problems and instruction for using pocket calculators. The calculator instruction is at the back of the book in Appendix A so that it may or may not be an integral part of the course depending on the teacher's wishes. However, even though this edition contains new lessons and problems, the primary objectives of this book have not changed.

This book is planned to help your students achieve the following four objectives:

1. Develop the ability to perform the fundamental operations in mathematics with a high degree of accuracy.
2. Produce results with reasonable speed through the use of practical shortcuts.
3. Establish, by repetition, the habit of forming clear numerals that are uniform in size.
4. Measure individual progress by use of a unique method of improvement scoring and charting of results in testing.

This book is designed to serve teachers who appreciate the value of a review in the fundamentals of mathematics. It can also be used profitably as a major text in a brief course or as a supplement to any mathematics textbook used in a longer course. This book is recommended for courses in clerical practice, distributive education, general business, and record keeping, as well as for elementary mathematics. It is suitable for young people who are preparing for service in the armed forces, for those who are in practical arts courses, for those preparing for a business occupation or for the business of everyday living, and for those planning to take State or Federal Civil Service entrance examinations.

With this book your students can attain accuracy and build speed in the fundamental operations of addition, subtraction, multiplication, and division. Emphasize the importance of neatness and of clear, uniform figures. The correct placement of digits and decimal points is stressed by example.

Measure Improvement. The method of measuring student progress in this book is distinctive. The exercises and tests used in the previous editions and in this edition were tried out with hundreds of students. They were tested to determine the number of problems that could be completed in a given length of time by an average student. The scores were then assigned so that the total possible score attainable on an exercise or test was approximately 200. The time was assigned so that the average student, without any preparation, could complete one half of an exercise or test and earn a score of 100, which is the basic score. The basic score of 100 is then subtracted from the total score. The difference is the improvement score because it indicates the improvement that the student has made through study and practice. As a student continues through the book, the improvement scores should rise. In order that the improvement scores may be compared, progress charts are provided on the inside of the front and back covers for the purpose of graphically recording these scores.

Build Enthusiasm. With this book and your guidance, any student who has the will can build mathematical skill. Start your class with a daily drill or test from this book and you will create a "get-down-to-business" attitude at the beginning of every class period. Let the students measure their own progress by means of the improvement scores. Then watch the interest of all your students skyrocket as they experience the joy of progress and achievement!

Acknowledgments. Many of the new lessons and several of the other changes in this edition are the result of suggestions made by many teachers who have used previous editions of this book in their classes. Their constructive assistance is hereby gratefully acknowledged.

Loyce C. Gossage

To the Student

When you have the will, you can develop your skill. With the exercises in this book, you can build a firm foundation in the fundamentals of mathematics so necessary for success in the business world and in the business of everyday living. Just as physical exercise is necessary to build a strong body, so mental exercise is necessary to build an alert mind. This book provides the opportunity for you to develop your ability to use figures accurately and to produce results with reasonable speed.

How to Use This Book. Each exercise in this book is followed by a timed test. Read the explanation and instructions at the top of each page. Practice each exercise until you can complete the exercise in the basic time shown. Then you should be ready to make a satisfactory record in the test given on the back of the sheet. The test is to be worked out only in the presence of your teacher. **No credit will be given for the test if the page is used prior to the session in which your teacher assigns it.**

Many errors in mathematics are due to careless writing of numerals, misplacement of digits and decimal points, and failure to form the habit of checking answers. You can avoid these pitfalls in mathematics if you write plain, clear numerals of uniform size, study the shortcuts, and follow the instructions of your teacher for checking your work.

Improvement Scores. In the tests you will not be graded on your total scores, but on your improvement scores. For each test, sufficient time is allowed for the average student to obtain a score of 100—the basic score—without previous study. The amount you earn above 100 is your improvement score. In the upper right-hand corner on each test, there is space for you to record your total score, and to enter the difference between the total score and the basic score which is your improvement score.

For scoring the tests, the value of each correct answer is shown after the instructions when each problem has the same weight. When values are printed at the right of the page, each answer in the row by the value has the same weight. When the value is printed at the bottom of a column and is followed by the word "each," each answer in that column has the value shown. If each answer has a different value, the value is given as near as possible to the space allowed for the answer. Ask your teacher to explain the scoring of any test if you have any question. **Do not remove the test sheet from the book until your teacher instructs you to do so.**

Progress Charts. On the inside front and back covers of this book, you have progress charts on which to keep an accurate record of your improvement scores. Use these charts for the improvement scores on the tests only, not on the practice exercises. The figures at the top and bottom of the charts are the test numbers; the figures at the sides of the charts indicate your possible improvement score for each test. To record your score, use a pen or a colored pencil and fill in the vertical bar for each numbered test until you reach the level that shows the score you have earned. Tall vertical bars indicate satisfactory progress. Short bars show that you need to spend more time on the practice exercises.

With preparation, practice, and perseverance you can build mathematical skill. Try!

Contents

Pretest   xi

Section 1 Whole Numbers

1 Reading and Writing Whole Numbers   1
2 Rounding Whole Numbers   3
3 Mental Estimation   5

Addition

4 Addition Facts   7
5 Addition of Numbers with More Than One Digit   9
6 Addition of Two-Digit Combinations   11
7 Combinations That Add to 10   13
8 Addition of Larger Numbers   15
9 Vertical and Horizontal Addition   17

Subtraction

10 Subtraction Facts   19
11 Checking Subtraction   21
12 Installment Buying Compared with Cash Buying   23
13 Bank Deposits and Checks   25
14 Customers' Accounts (Accounts Receivable) 27
15 Application Problems   29

Multiplication

16 Multiplication, 196 Key Combinations   31
17 Multiplication with One-Digit Multipliers 33
18 Multiplication with Two- and Three-Digit Multipliers   35
19 Multiplication with 0 in Multiplicand and in Multiplier   37
20 Multiplication by 10 and by Multiples of 10   39
21 Application Problems   41

Division

22 Division Facts   43
23 Division with Small Divisors   45
24 Long Division   47
25 Zeros in the Quotient   49
26 Finding Averages   51
27 Application Problems   53

Cumulative Review 1   55

Section 2 Equations

28 Signed Numbers—Addition and Multiplication   59
29 Signed Numbers—Division and Subtraction   61
30 Solving Equations   63
31 Multiplying and Collecting Terms in Equations   65
32 Order of Operations   67
33 Parentheses in Equations   69
34 Algebraic Expressions   71
35 Application Problems   73

Cumulative Review 2   75

Section 3 Fractions

Denominators

36 Common Denominator	79	39 Lowest Common Denominator	85
37 Lowest Terms of Common Fractions	81	40 Estimation with Fractions	87
38 Improper Fractions and Mixed Numbers	83		

Addition and Subtraction

41 Addition of Common Fractions	89	43 Subtraction of Common Fractions	93
42 Addition of Mixed Numbers	91	44 Subtraction of Mixed Numbers	95

Multiplication and Division

45 Multiplication of Common Fractions	97	49 Division of Mixed Numbers	105
46 Multiplication of Mixed Numbers	99	50 Simplification of Complex Fractions	107
47 Multiplication of Whole Numbers by Common Fractions	101	51 Review of Fractions	109
48 Division of Common Fractions	103	52 Application Problems	111

Cumulative Review 3   113

Section 4 Decimal Fractions and Aliquot Parts

53 Reading and Rounding Decimals	117	58 Figuring Gross Pay	127
54 Equivalent Decimal and Common Fractions	119	59 Division of Decimals	129
55 Decimal Equivalents	121	60 Shortcut Division by 10 and by Multiples of 10	131
56 Estimation with Decimals	123	61 Aliquot Parts	133
57 Placement of the Decimal Point	125		

Cumulative Review 4   135

Section 5 Percentage and Selected Business Topics

Percents

62 Changing Percents to Decimals and Fractions	139	64 Estimation with Percents	143
63 Changing Decimals and Fractions to Percents	141	65 Percentage Problems	145

Commissions

66 Commission on Sales	147	67 Commission on Purchases	149

Discounts

68 Trade Discounts	151	70 Cash Discounts	155
69 Determining Due Dates	153	71 Discounts on Invoices	157

Interest

72 Simple Interest Formula	159	73 Compound Interest	161

Cumulative Review 5   163

Section 6 Measurement

United States Measurement

74 U. S. Measurements 167	75 Denominate Numbers 169

Metric Measurement

76 Vocabulary of Metric System of Measurement 171	79 Converting U. S. Measurements to Metric Measurements 177
77 Metric Measurements 173	80 Converting Metric Measurements to U. S. Measurements 179
78 Metric Denominates 175	

Cumulative Review 6 181

POSTTEST 183

TABLE I 185
U.S. Units of Measure and Their Approximate Metric Equivalents

TABLE II 186
Metric Units of Measure and Their Approximate U.S. Equivalents

APPENDIX Pocket Calculators

A1 Calculator Operations 187	A3 Calculator Fractions and Constants 191
A2 Calculator Memory 189	A4 Calculator Cutoff, Overflow, and Percent 193

Cumulative Review A 195

Pretest

Name _____ Score _____

Solve these problems. Place each answer in the appropriate space on the right. Show common fractions in lowest terms. (Score 2 points for each correct answer to Problems 1 through 20 and 3 points for each correct answer to Problems 21 through 40.)

Add:

1. 630
   285
   519
   326
   572

2. $5.20 + 2.87 + 0.09 + 4.06$

3. $\dfrac{7}{8}$
   $\dfrac{9}{16}$

4. $17\dfrac{11}{12}$
   $9\dfrac{3}{4}$

Subtract:

5. 8,700
   3,986

6. $8.35 - 4.376 =$

7. $\dfrac{7}{12} - \dfrac{1}{4} =$

8. $9\dfrac{1}{10}$
   $4\dfrac{2}{5}$

Multiply:

9. 784
   605

10. $\dfrac{7}{8} \times 44 =$

11. $\dfrac{7}{9} \times \dfrac{8}{13} =$

12. $17\dfrac{2}{3}$
    $6\dfrac{3}{5}$

13. $80 \times 0.80 =$

14. $3.2 \times 0.9 =$

Divide:

15. $954 \div 9 =$

16. $0.806 \div 0.008 =$

17. $18 \div \dfrac{5}{6} =$

18. $\dfrac{5}{6} \div \dfrac{2}{3} =$

19. $8\dfrac{3}{8} \div 17 =$

20. $5\dfrac{1}{8} \div 10\dfrac{1}{4} =$

1. _____
2. _____
3. _____
4. _____
5. _____
6. _____
7. _____
8. _____
9. _____
10. _____
11. _____
12. _____
13. _____
14. _____
15. _____
16. _____
17. _____
18. _____
19. _____
20. _____

Pretest

Write each of the following as a percent:

21. 0.0875

22. $\frac{3}{4}$

23. 3

Write each of the following as a common fraction:

24. 50%

25. $16\frac{2}{3}\%$

26. $\frac{1}{4}\%$

Write each of the following as a decimal fraction:

27. $1\frac{1}{4}$

28. $37\frac{1}{2}\%$

29. Find the average of these daily attendance numbers:
    1,489   1,587   1,628   1,406   1,531   1,593

30. Solve this equation: $4(x + 3) - 3(x - 5) = 145$

Solve each of the following:

31. $12\frac{1}{2}\%$ of 424 = __?__

32. 36 = 6% of __?__

33. 24 = __?__% of 30

34. Add: 4 hr 32 min
            8 hr 45 min

35. 32 km = __?__ m

36. An agent sold a client's shipment of fruit and collected $2,840. The charges were $79.50 for freight and 6% commission for selling. How much should the agent send to the client?

37. On March 16, a dealer listed terms of 2/10, n/30 on an invoice for $1,800 of merchandise. How much should the customer pay on March 25?

38. Find the net price for merchandise listed on an invoice at $900 less trade discounts of 25% and 10%.

39. How much is the simple interest on $900 at 14% for nine months?

40. A student deposited $800 in an account that pays 12% interest compounded monthly. Find the compound amount at the end of four months.

LESSON 1

Reading and Writing Whole Numbers

To read or write a whole number, arrange its digits in groups of three digits each. Starting with the first whole digit on the right, insert a comma to the left of each three-digit group. Notice the name of each group as shown in the chart on the right. Each three-digit group is read as if the group itself were a three-digit number.

The numbers in the chart are read as follows:

A. Five hundred sixty-one. Note on the chart that 5 is in the "hundreds" column, 6 is in the "tens" column, and 1 is in the "ones" column.
B. Seven hundred forty-six thousand, five hundred seventeen.
C. Eight million, three hundred twelve thousand, forty-eight.
D. Forty-eight billion, two hundred fifty-two million, six hundred thirty-one thousand, two hundred thirty-five.

	Trillion			Billion			Million			Thousand			Units		
	Hundreds	Tens	Ones	Hundreds	Tens	Ones	Hundreds	Tens	Ones	Hundreds	Tens	Ones	Hundreds	Tens	Ones
A													5	6	1
B										7	4	6	5	1	7
C									8	3	1	2	0	4	8
D						4	8	2	5	2	6	3	1	2	3	5
E	8	3	2	0	7	4	0	0	5	0	0	0	0	0	9

E. Eight hundred thirty-two trillion, seventy-four billion, five million, nine. In reading **E,** notice that the group representing thousands is not mentioned because the zeros in this group indicate "no thousands."

EXERCISE 1/Basic Time—$3\frac{1}{2}$ Minutes

Estimated time to obtain a basic score of 100. The basic time for each exercise in this book is the estimated time to earn a basic score of 100. It should be possible to complete any exercise in this book in approximately twice the basic time.

Write the following numbers in neat, legible numerals with commas inserted where applicable: (10 each)

1. Seventeen
2. Fifty-four
3. Eight hundred seventy-five
4. Seven hundred eighty
5. Four thousand, two hundred seventy-five
6. Twelve thousand, two hundred eighty-seven
7. Nine hundred sixty-two thousand, two hundred six
8. Three million, four hundred forty thousand, one hundred ninety-one
9. Two billion, four million, three thousand, five
10. Eight trillion, six billion, four million, two hundred
11. Seventy-eight million, sixty-four thousand, nine hundred forty-five
12. Six thousand, fifty
13. Forty-two thousand, seven
14. Eighty-nine billion, five hundred thousand, nine hundred eighty
15. Seven hundred fifty-nine thousand, three hundred twenty-four
16. Four hundred fifty-three billion, nine hundred three million, six hundred ninety-two thousand, ninety
17. Seventy-five trillion, sixty-two million, twelve thousand
18. Five hundred eighty million, nine hundred five thousand, six hundred eleven
19. Sixty-three billion, five hundred eighty million, fifty-eight thousand, ninety-six
20. Three hundred seventy-six billion, nine hundred seventy-one thousand, seven hundred forty

Name	Total Score
Date	Basic Score 100
Basic Time—3 Minutes	Improvement Score

You should be able to score 100 on this test in 3 minutes. The amount you earn above 100 is your improvement score. The improvement scores for Tests 1 through 40 should be recorded on the Progress Chart on the inside of the front cover, and the improvement scores for Tests 41 through 80 should be recorded on the inside of the back cover. Record this test as the first vertical bar on the inside front cover for Test 1.

Write these numbers in neat, legible numerals and insert commas where applicable: (10 each)

1. Fourteen
2. Eighty-seven
3. Six hundred twenty-seven
4. One hundred thirty-one
5. Eleven thousand, three hundred fourteen
6. Eighty-five thousand, eighty
7. Two hundred nine thousand, five
8. Four million, fifty-five thousand, two hundred forty-eight
9. Seventy-six million, seven hundred ninety-nine thousand, two hundred twenty-seven
10. Four hundred one million, eight thousand, sixty
11. Seventy billion, one hundred seventy-three million, two hundred one thousand, one hundred
12. Fifty-four billion, five hundred two million, four hundred eight thousand, five hundred nine
13. Five hundred sixty-one billion, eight hundred eight million, sixty-five thousand, four
14. Nineteen trillion, five hundred billion, eight hundred million, seven thousand
15. Sixty trillion, seven hundred sixty-two million, five hundred eighteen
16. Forty thousand, five hundred
17. Two billion, one hundred fifty-nine thousand, three hundred fifty-eight
18. Three hundred billion, five hundred sixty-eight million, thirty
19. Seventy-nine billion, two hundred fifty million, two hundred fifty-six thousand, one hundred forty-eight
20. Eight million five

1. _____
2. _____
3. _____
4. _____
5. _____
6. _____
7. _____
8. _____
9. _____
10. _____
11. _____
12. _____
13. _____
14. _____
15. _____
16. _____
17. _____
18. _____
19. _____
20. _____

LESSON 2: Rounding Whole Numbers

To round a number, follow these rules:
1. Underline the digit in the specified place. This is the *place* digit. The digit to the immediate right of the place digit is the *test* digit.
2. If the test digit is 5 or larger, add 1 to the place digit and substitute zeros for all digits to its right.
3. If the test digit is 4 or smaller, substitute zeros for it and all digits to the right.

In rounding 4,765,832 to its nearest hundred thousand, the place digit is 7 and the test digit is 6.

Place digit ↓
4,7̲65,832
Test digit ↑

As example C illustrates, adding 1 to the place digit can necessitate increasing the digit to its left.

Example A: Round 4,756,832 to nearest hundred thousand.
4,7̲65,832   Test digit (6) is five or larger.
+ 1   Place digit (7) is increased by 1
4,800,000   and zeros are substituted to the right.

Example B: Round 75,342 to nearest thousand.
75,342   Test digit (3) is 4 or smaller.
             Place digit (5) remains unchanged,
75,000   and zeros are substituted to the right.

Example C: Round 529,671,364 to nearest million.
529,̲671,364 Test digit (6) is 5 or larger.
+ 1   Place digit (9) is increased by 1,
530,000,000 and zeros are substituted in number.

EXERCISE 2 / Basic Time – 3 Minutes

Estimated time to obtain a basic score of 100.
Round the following numbers to the places indicated: (10 each)

1. 264 to nearest ten
2. 267 to nearest ten
3. 6,400 to nearest thousand
4. 2,060 to nearest hundred
5. 21,220 to nearest hundred
6. 782,650 to nearest thousand
7. 518,660 to nearest thousand
8. 204,746 to nearest ten thousand
9. 207,614 to nearest ten thousand
10. 7,880,000 to nearest hundred thousand
11. 4,460,132,158 to nearest hundred million
12. 72,059,800,000 to nearest million
13. 103,097,356,226 to nearest ten million
14. 1,026,844,671,486 to nearest billion
15. 440,199,734,687,584 to nearest hundred billion
16. 105,088,637,430,398 to nearest hundred million
17. 498,497,796,938,350 to nearest ten trillion
18. 67,950,701,909,637 to nearest trillion
19. 589,549,980,000 to nearest billion
20. 360,306,077,500 to nearest ten million

TEST 2

Name _____
Date _____
Basic Time–3 Minutes

Total Score _____
Basic Score __100__
Improvement Score _____

Round the following numbers to the places indicated: (10 each)

1. 2,008 to nearest ten
2. 2,024 to nearest ten
3. 2,710 to nearest hundred
4. 3,761 to nearest hundred
5. 89,108 to nearest thousand
6. 472,837 to nearest thousand
7. 313,866,334 to nearest million
8. 734,196,855,447 to nearest ten million
9. 13,588,588 to nearest ten thousand
10. 53,716,240 to nearest ten million
11. 190,897,794,295 to nearest billion
12. 36,823,462,770 to nearest hundred thousand
13. 211,867,752,000 to nearest hundred million
14. 10,756,734,204,596 to nearest trillion
15. 6,041,924,317,695 to nearest hundred thousand
16. 1,239,842,763,521,400 to nearest hundred trillion
17. 6,907,997,702,393,762 to nearest ten billion
18. 52,405,607,322,520 to nearest hundred billion
19. 718,428,674,337,579 to nearest ten trillion
20. 5,799,679,750,569,459 to nearest billion

1. _____
2. _____
3. _____
4. _____
5. _____
6. _____
7. _____
8. _____
9. _____
10. _____
11. _____
12. _____
13. _____
14. _____
15. _____
16. _____
17. _____
18. _____
19. _____
20. _____

© Copyright South-Western Publishing Co.

LESSON 3

Mental Estimation

The purpose of this exercise is to help you acquire confidence in computing answers. An *estimate* is an approximation of the exact answer to a computation. Rounded numbers are used to make quick mental calculations. Understanding and using mental estimation will help you to determine whether your answers "make sense." Answers that don't make sense are usually wrong.

Example A: The problem 8,250 − 5,720 can be rounded to 8,000 − 6,000, which is 2,000. Therefore, the exact answer is near 2,000 and must have 4 digits.

Example B: 3,815 + 670 + 2,050 rounds to 4,000 + 1,000 + 2,000 = 7,000.

Because the 3,815 and 670 were rounded up more than the 2,050 was rounded down, the estimate of 7,000 is high and should be decreased. Therefore, the estimate may be shown as 7,000−.

Example C: 8,315 ÷ 36 rounds to 8,000 ÷ 40, giving an estimate of 200. As the 8,000 is too small and the 40 is too large, the estimate is low and should be increased. The estimate may be shown as 200+.

EXERCISE 3/Basic Time−2 Minutes

Estimated time to obtain a basic score of 100.

Look at each problem. Do not show the exact answer or an estimate. Show the number of digits there should be in each answer. (10 each)

1. 262 + 65 + 628   1. _____ Digits   2. 431 + 87 + 718   2. _____ Digits
3. 6,000 − 4,362   3. _____ Digits   4. 5,000 − 4,780   4. _____ Digits
5. 41 × 52   5. _____ Digits   6. 381 × 7   6. _____ Digits
7. 82,570 ÷ 8   7. _____ Digits   8. 7,324 ÷ 74   8. _____ Digits

For each of the following, show an estimate. Then, show whether your estimate should be increased or decreased by circling the appropriate symbol. Do not adjust your estimate or compute the exact answer. (10 each)

9. 39 × 48   9. Estimate _____ + −   10. 516 × 67   10. Estimate _____ + −

11. 4,203 ÷ 8   11. Estimate _____ + −   12. 32,657 ÷ 37   12. Estimate _____ + −

13. 5,234   14. 462
      976   327
    +2,641   +285
    ******   ****
                        13. Estimate _____ + −   14. Estimate _____ + −

15. 3,160   16. 75,850
    −1,920   −56,176
    ******   *******
                        15. Estimate _____ + −   16. Estimate _____ + −

17.   437   18.   280
    ×   8   ×73
     ****   ***
                        17. Estimate _____ + −   18. Estimate _____ + −

19.   *****   20.   *****
    8)3,462   29)68,435
                        19. Estimate _____ + −   20. Estimate _____ + −

TEST 3

Name _____ Total Score _____
Date _____ Basic Score __100__
Basic Time–2 Minutes Improvement Score _____

Look at each problem. Do not show the exact answer or an estimate. Show the number of digits there should be in each answer. (10 each)

1. 625 + 57 + 281   1. _____ Digits   2. 515 + 870 + 682   2. _____ Digits
3. 7,000 − 3,470   3. _____ Digits   4. 27,000 − 8,750   4. _____ Digits
5. 57 × 64   5. _____ Digits   6. 410 × 82   6. _____ Digits
7. 9,461 ÷ 9   7. _____ Digits   8. 76,120 ÷ 73   8. _____ Digits

Show an estimate. Then show whether your estimate should be increased or decreased by circling the appropriate symbol. Do not adjust your estimate or compute the exact answer. (10 each)

9. 46 × 52   9. Estimate _____ + −   10. 615 × 8   10. Estimate _____ + −
11. 5,630 ÷ 9   11. Estimate _____ + −   12. 34,740 ÷ 72   12. Estimate _____ + −

13. 6,523   14. 7,260
      861   931
    +3,170   +6,585
    *****   13. Estimate _____ + −   *****   14. Estimate _____ + −

15. 4,735   16. 84,620
    −1,872   −57,900
    *****   15. Estimate _____ + −   ******   16. Estimate _____ + −

17.   741   18.   383
      × 9   ×82
      ***   17. Estimate _____ + −   ****   18. Estimate _____ + −

       *****   ******
19. 8)5,630   19. Estimate _____ + −   20. 38)92,536   20. Estimate _____ + −

LESSON 4
Addition Facts

Addition is combining two or more numbers (*addends*) to obtain an answer (*sum* or *total*).

The secret of speed and accuracy in addition is recognizing combinations of numbers as totals, just as combinations of letters are recognized as words. Using 0, 1, 2, 3, 4, 5, 6, 7, 8, and 9, there are 100 combinations of two numbers. These combinations are shown below. *Study them thoroughly, left to right, up and down, and on the diagonal*, to learn all the combinations of these numbers. Although this exercise is a very simple one, it will form part of your mathematical background for future work.

The diagonal rows show the combinations that form the totals from 0 to 18. The combinations on the diagonal row for 10 are very important to learn, because they will be in another exercise.

Practice this exercise until you can complete it in less than one minute.

EXERCISE 4/Basic Time—$\frac{3}{4}$ Minute
Estimated time to obtain a basic score of 100.

Add: (2 each)

	a	b	c	d	e	f	g	h	i	j	
	0	1	2	3	4	5	6	7	8	9	
1.	0+0	1+0	2+0	3+0	4+0	5+0	6+0	7+0	8+0	9+0	10
2.	0+1	1+1	2+1	3+1	4+1	5+1	6+1	7+1	8+1	9+1	11
3.	0+2	1+2	2+2	3+2	4+2	5+2	6+2	7+2	8+2	9+2	12
4.	0+3	1+3	2+3	3+3	4+3	5+3	6+3	7+3	8+3	9+3	13
5.	0+4	1+4	2+4	3+4	4+4	5+4	6+4	7+4	8+4	9+4	14
6.	0+5	1+5	2+5	3+5	4+5	5+5	6+5	7+5	8+5	9+5	15
7.	0+6	1+6	2+6	3+6	4+6	5+6	6+6	7+6	8+6	9+6	16
8.	0+7	1+7	2+7	3+7	4+7	5+7	6+7	7+7	8+7	9+7	17
9.	0+8	1+8	2+8	3+8	4+8	5+8	6+8	7+8	8+8	9+8	18
10.	0+9	1+9	2+9	3+9	4+9	5+9	6+9	7+9	8+9	9+9	

© Copyright South-Western Publishing Co.

TEST 4

Name _____ Total Score _____
Date _____ Basic Score _____
Basic Time – $\tfrac{3}{4}$ Minute   Improvement Score _____

Add: (2 each)

	a	b	c	d	e	f	g	h	i	j
1.	9 8	8 6	7 9	6 3	3 8	2 4	8 3	9 5	3 5	9 2
2.	3 2	5 8	2 3	4 8	9 9	1 1	8 0	4 7	0 9	4 1
3.	1 0	8 8	6 6	8 4	3 6	1 5	3 0	3 3	8 5	5 3
4.	9 6	4 0	2 6	7 7	5 4	6 0	0 8	2 1	9 4	5 0
5.	0 7	7 0	7 8	6 7	7 1	2 5	2 8	8 9	9 7	4 5
6.	8 7	0 3	4 9	2 2	9 1	5 9	9 3	0 2	1 7	1 3
7.	3 1	6 5	7 4	1 6	2 0	1 9	6 9	7 6	6 2	8 2
8.	6 8	5 1	2 7	5 6	3 4	6 4	0 0	7 2	8 1	3 9
9.	4 2	0 5	4 3	7 3	1 8	4 6	3 7	5 5	0 6	6 1
10.	5 2	1 4	7 5	4 4	1 2	0 4	5 7	2 9	0 1	9 0

© Copyright South-Western Publishing Co.

LESSON 5
Addition of Numbers with More Than One Digit

The Arabic numerals 1-9 (and sometimes 0) are referred to as *digits*. This drill combines the addition of one-digit and two-digit numbers. When the digits in the right column total more than 9, a digit must be carried over to the next column on the left. It is helpful to write this digit over the left column so it will not be forgotten.

The example to the right shows the carrying of the digit 1. When adding 26 and 8, the digit 1 is carried over to the "tens" column from the total of 14 in the "ones" column. Before actually writing the answers, add these numbers mentally for practice in addition.

```
 1
26
 8
──
34
```

EXERCISE 5/Basic Time—1 Minute
Estimated time to obtain a basic score of 100.

Add: (2 each)

	a	b	c	d	e	f	g	h	i	j
1.	11 _2	15 _9	18 _6	12 _5	17 _8	14 _7	19 _4	16 _1	13 _6	18 _9
2.	23 _4	2 26	21 _4	2 25	28 _9	7 24	29 _5	6 22	26 _8	9 27
3.	3 37	8 34	7 32	5 39	3 31	2 35	9 38	5 33	3 36	7 38
4.	41 _5	6 48	42 _8	2 47	44 _8	7 49	46 _4	6 43	48 _6	9 46
5.	6 56	52 _9	4 54	59 _8	7 53	4 55	58 _5	51 _6	9 58	53 _9
6.	62 _5	66 _6	7 61	4 67	69 _9	64 _6	3 68	5 65	64 _5	68 _6
7.	5 75	72 _2	76 _7	6 73	4 78	77 _5	88 _5	8 86	9 83	6 75
8.	87 _6	7 85	4 92	9 96	99 _4	93 _9	97 _7	9 98	5 98	7 96
9.	3 13	27 _9	36 _7	45 _3	3 57	8 63	8 89	95 _8	38 _8	7 26
10.	23 _6	9 34	42 _7	8 59	69 _4	7 77	97 _5	9 95	9 87	98 _9

© Copyright South-Western Publishing Co.

TEST 5

Name _____ Total Score _____
Date _____   Basic Score _100_
Basic Time–1 Minute   Improvement Score _____

Add: (2 each)

	a	b	c	d	e	f	g	h	i	j
1.	11 _9_	26 _7_	32 _6_	47 _8_	53 _6_	64 _5_	79 _3_	85 _1_	95 _2_	37 _5_
2.	4 _14_	2 _21_	3 _39_	2 _44_	8 _55_	7 _68_	3 _71_	6 _81_	9 _89_	8 _46_
3.	25 _3_	8 _31_	49 _7_	9 _58_	63 _4_	3 _77_	12 _9_	5 _28_	39 _6_	7 _38_
4.	5 _46_	51 _4_	3 _65_	17 _8_	9 _24_	38 _9_	4 _43_	57 _6_	9 _75_	47 _6_
5.	29 _5_	33 _9_	5 _45_	8 _16_	22 _7_	36 _4_	4 _16_	6 _27_	64 _8_	5 _37_
6.	6 _23_	6 _34_	42 _7_	59 _4_	9 _69_	4 _74_	87 _2_	98 _8_	3 _98_	5 _59_
7.	48 _6_	54 _2_	7 _67_	78 _7_	83 _4_	5 _91_	41 _8_	52 _8_	4 _37_	28 _6_
8.	4 _73_	5 _89_	97 _8_	6 _56_	8 _66_	76 _7_	6 _82_	7 _94_	49 _5_	8 _39_
9.	72 _4_	86 _9_	93 _3_	5 _75_	5 _88_	7 _99_	84 _9_	96 _6_	57 _9_	6 _37_
10.	7 _95_	18 _9_	9 _35_	14 _3_	5 _37_	61 _5_	5 _13_	62 _9_	5 _36_	9 _47_

LESSON 6
Addition of Two-Digit Combinations

If you have thoroughly mastered the simple combinations, you will have little difficulty adding two two-digit numbers.

The *reverse-order check* is a popular method of checking addition accuracy. The sum of 40 + 50 is 90; conversely, 50 + 40 is 90. If you add numbers down, check the accuracy of the sum by adding the numbers up.

Practice these problems mentally before writing the answers.

EXERCISE 6/Basic Time–2 Minutes
Estimated time to obtain a basic score of 100.

Add: (2 each)

	a	b	c	d	e	f	g	h	i	j
1.	38 21	58 32	45 34	32 25	45 54	32 28	65 27	83 15	64 28	39 29
2.	28 49	87 46	33 86	29 45	34 28	93 19	47 68	25 39	23 88	74 23
3.	68 49	56 72	64<>28	39 42	37 63	55 83	74 16	19 47	65 87	98 43
4.	35 22	59 76	86 24	55 99	24 13	82 74	55 29	82 36	56 49	39 26
5.	97 14	26 53	27 84	53 28	28 73	47 23	84 29	31 45	75 78	93 28
6.	78 25	37 69	88 14	73 69	84 28	62 47	92 17	73 16	29 33	34 69
7.	59 67	32 49	55 63	87 93	48 26	56 97	45 57	63 98	37 69	84 59
8.	23 89	76 25	93 25	62 54	83 17	75 29	36 83	24 47	75 98	64 36
9.	65 28	24 57	28 16	74 28	68 93	95 83	49 87	63 26	62 79	65 97
10.	54 78	69 23	54 28	49 83	72 96	47 84	26 75	38 44	84 29	58 46

© Copyright South-Western Publishing Co.

TEST 6

Name _____ Total Score _____
Date _____ Basic Score 100
Basic Time–2 Minutes Improvement Score _____

Add: (2 each)

	a	b	c	d	e	f	g	h	i	j
1.	39 83	57 37	47 29	63 34	42 93	34 47	65 25	83 24	85 64	57 69
2.	28 56	23 46	32 86	34 45	25 24	54 15	28 68	25 39	96 69	28 37
3.	28 49	49 71	68 55	39 24	37 82	59 58	74 82	19 74	79 59	82 77
4.	37 26	52 29	83 53	25 99	42 13	62 71	83 29	16 36	85 63	57 64
5.	22 53	76 84	25 28	28 62	47 92	84 73	36 95	68 39	84 73	96 68
6.	37 55	82 87	75 48	84 53	73 49	26 15	29 16	45 82	24 88	47 29
7.	82 67	67 68	15 57	28 93	47 28	78 34	93 67	27 43	62 98	25 89
8.	43 56	94 26	87 49	27 36	83 29	78 68	26 84	95 97	87 49	38 57
9.	82 84	75 68	23 77	49 59	68 73	28 29	15 26	53 74	84 59	67 46
10.	73 28	26 84	19 27	34 26	83 15	73 68	92 55	73 26	79 82	87 29

12 © Copyright South-Western Publishing Co.

LESSON 7: Combinations that Add to 10

Up to now, the problems have been limited to the addition of two numbers. When adding more than two numbers, there are instances when two digits that add to 10 come together. Form the habit of recognizing these combinations as 10 and adding them in a single operation. Grouping combinations that add to 10 will greatly increase both your accuracy and speed.

In this exercise, each problem contains at least one combination of 10. Be extremely careful to avoid overlooking the numerals that are not included in the combination. At first, indicating these combinations as illustrated in Example 1 will be helpful.

In the first four drills below, combinations of 10 are connected by lines. In the fifth drill, you are to connect the combinations by lines before adding.

Example 1:

```
     8
10 ( 6
     2 ) 10
     4
     7
    27
```

EXERCISE 7 / Basic Time – 2½ Minutes

Estimated time to obtain a basic score of 100.

Add:

	a	b	c	d	e	f	g	h	i	j	k	l
1.	6 4 7	8 9 1	9 4 6	8 2 7	7 3 7	5 5 8	5 3 5	3 9 7	1 6 9	6 8 4	8 7 2	5 8 5

(2)

	a	b	c	d	e	f	g	h	i	j	k
2.	3 7 4 1 4 6	6 4 2 8 4 2	1 8 9 1 5 5	4 6 2 7 8 2	3 9 6 4 3 7	2 8 4 6 7 9	3 6 7 2 8 3	2 9 7 1 4 6	3 6 4 8 3 2	7 6 3 8 6 2	7 5 2 5 8 4	3 6 7 4 2 5

(4)

	a	b	c	d	e	f	g	h
3.	33 67 42	32 26 84	74 32 28	39 81 76	47 57 63	28 41 82	36 81 74	67 36 43

(4)

	a	b	c	d	e	f	g	h
4.	82 78 32 18 96 54	47 63 82 28 99 51	74 36 48 87 23 45	18 92 47 63 93 27	75 48 35 62 53 48	83 26 77 47 32 63	32 79 21 64 86 21	99 81 47 23 64 86

(8)

Connect the combinations of 10 by lines and add:

	a	b	c	d	e	f	g	h
5.	64 78 32	72 39 31	72 38 25	87 29 43	26 17 84	38 29 72	41 59 65	24 56 87

(4)

TEST 7

Name _____ Total Score _____
Date _____ Basic Score 100
Basic Time – 2½ Minutes Improvement Score _____

Add:

	a	b	c	d	e	f	g	h	i	j	k	l
1.	6 4 8	6 2 8	9 3 7	9 1 8	5 5 7	5 6 4	1 8 9	6 9 4	5 9 5	3 2 7	8 9 2	2 7 8 (2)

	a	b	c	d	e	f	g	h	i	j	k	l
2.	3 7 2 8 2 5	7 4 6 7 8 2	9 1 8 2 4 8	7 1 9 2 8 4	4 9 8 2 4 6	8 2 4 6 3 8	9 1 4 7 6 3	5 4 4 6 6 2	3 6 4 9 8 1	2 6 4 8 9 2	3 7 4 6 8 4	6 8 3 5 7 5 (4)

	a	b	c	d	e	f	g	h
3.	54 56 27	38 71 49	84 29 31	75 28 82	92 44 18	27 49 83	73 56 34	29 84 11 (4)

	a	b	c	d	e	f	g	h
4.	16 74 37 29 81 52	36 44 68 82 26 34	83 27 49 62 88 23	37 63 44 29 81 67	64 26 82 51 59 35	36 27 74 23 85 35	78 29 32 81 56 64	78 22 36 75 66 45 (8)

Connect the combinations of 10 by lines and add:

	a	b	c	d	e	f	g	h
5.	34 49 61	83 37 72	65 45 82	74 56 58	29 41 87	65 25 43	72 47 68	39 81 76 (4)

LESSON 8: Addition of Larger Numbers

The addition of numbers with three digits or more is presented here to develop greater speed and accuracy. Form the habit of making legible numbers and keeping digits in a straight column. Insert commas where necessary.

In adding digits that total over 9, be sure to carry a digit over to the next column on the left.

In the example at the right, the digit 1 is carried over from the far right column to the middle column. Another 1 is carried over from the middle column to the left column.

It helps to write at the top of the column the digit which is carried to that column.

Example:
11
234
587
821

EXERCISE 8/Basic Time – 2 Minutes
Estimated time to obtain a basic score of 100.

Add:

	a	b	c	d	e	f	
1.	345 698	254 736	829 847	378 690	576 692	925 381	(4)
2.	748 642	293 718	786 694	267 545	493 386	254 255	(4)
3.	843 927	298 354	674 737	235 872	368 645	758 458	(4)

	a	b	c	d	
4.	642,865 921,565	784,228 843,627	372,487 687,769	478,695 341,217	(8)

	a	b	c	
5.	742,618,726 427,864,415	427,893,876 738,264,696	924,647,826 376,256,878	(12)
6.	423,876,989 839,213,469	476,827,698 928,022,809	247,692,385 786,427,820	(12)
7.	294,398 748,696	249,387 389,827	831,476 978,937	(8)

TEST 8

Name _____ Total Score _____
Date _____ Basic Score _100_
Basic Time–2 Minutes Improvement Score _____

Add:

	a	b	c	d	e	f	
1.	786 394	928 782	475 698	378 254	689 836	546 926	(4)
2.	375 269	543 786	834 729	547 269	786 293	548 687	(4)
3.	287 465	387 298	546 589	728 569	834 785	699 847	(4)

	a	b	c	d	
4.	683,945 821,596	382,769 737,899	473,586 469,218	592,768 419,826	(8)

	a	b	c	
5.	725,678,296 854,927,705	537,698,276 654,276,987	547,692,859 382,769,848	(12)
6.	738,925,467 426,178,698	293,847,693 542,769,827	786,925,478 354,769,845	(12)
7.	392,844 547,276	693,295 248,708	476,698 781,788	(8)

16 © Copyright South-Western Publishing Co.

LESSON 9
Vertical and Horizontal Addition

Frequently in business, groups of numbers must be added both horizontally and vertically. Records of daily cash receipts and daily expense account summaries are examples of this type of addition.

An excellent means of checking the accuracy of such addition is to verify that the grand total of the sums of the vertical columns equals the total of the sums of the horizontal rows, as shown in the example. Notice that the total of the sums of the horizontal rows (9, 17, and 19) is 45, and the total of the sums of the vertical columns (22, 16, and 7) is also 45. A single error at any point, of course, makes this equality impossible.

Be very careful when adding horizontally to add together only those digits in the same position in each column of numbers.

Example:
$$5 + 3 + 1 = 9$$
$$8 + 7 + 2 = 17$$
$$\underline{9 + 6 + 4} = \underline{19}$$
$$22 + 16 + 7 = 45$$

EXERCISE 9 / Basic Time—15 Minutes
Estimated time to obtain a basic score of 100.

Add horizontally and vertically: (14 points for No. 14; 4 points for other correct answers)

1.	375	24	8	326	493	7	65	
2.	47	45	549	9	380	74	874	
3.	681	394	27	7	745	88	68	
4.	6	27	683	548	29	926	29	
5.	927	6	57	9	574	54	385	
6.	63	983	8	28	689	75	9	
	7.	8.	9.	10.	11.	12.	13.	14.

Add horizontally and vertically: (20 points for No. 34; 6 points for other correct answers)

	CASH RECEIPTS							
	Route Number	Mon.	Tues.	Wed.	Thurs.	Fri.	Weekly Total
15.	1	45 11	50 19	19 72	60 44	70 12	
16.	2	53 10	40 12	32 19	60 12	79 42	
17.	3	39 12	50 26	30 91	87 21	87 63	
18.	4	42 40	24 04	36 40	89 42	96 19	
19.	5	38 67	36 60	43 90	77 09	86 41	
20.	6	29 05	47 29	40 41	87 90	91 11	
21.	7	30 17	37 02	50 04	97 82	97 36	
22.	8	51 42	49 11	48 19	98 64	92 42	
23.	9	19 09	38 05	49 71	89 46	68 50	
24.	10	35 04	60 51	30 65	74 11	72 76	
25.	11	68 29	37 10	49 11	79 32	84 62	
26.	12	32 22	43 32	50 09	84 47	93 99	
27.	13	29 40	32 34	56 14	91 10	92 84	
28.	14	41 13	73 40	48 39	88 05	85 01	
	Daily Total	29.	30.	31.	32.	33.	34.

TEST 9

Name _____ Total Score _____
Date _____ Basic Score __100__
Basic Time–15 Minutes Improvement Score _____

Add horizontally and vertically: (10 points for Nos. 16 and 32; 2 points for other correct answers)

1.	7	8	5	6	9	4	7	17.	5	9	4	3	8	2	8
2.	4	9	6	5	8	6	9	18.	9	8	7	4	6	5	3
3.	9	4	6	7	3	5	8	19.	7	4	9	6	8	2	5
4.	3	8	4	3	6	7	4	20.	6	7	3	9	7	8	9
5.	6	3	9	4	5	8	9	21.	3	7	5	7	5	6	7
6.	8	5	3	9	3	9	5	22.	8	5	8	5	9	7	4
7.	5	7	8	4	7	3	6	23.	4	6	7	8	4	9	6
8.	2	6	7	8	4	5	4	24.	9	8	6	9	7	6	8

9. 10. 11. 12. 13. 14. 15. 16.   25. 26. 27. 28. 29. 30. 31. 32.

Some business firms offer their salespeople bonuses, usually in the form of extra cash commissions, on all sales exceeding set amounts.

Four salespeople earned bonuses for additional sales as shown below. Find the total bonuses paid by the firm during each month, the total bonus earned by each salesperson during the year, and the grand total of bonuses paid during the year. (24 points for No. 49; 6 points for other correct answers)

	MONTH	BLANCO	CRUZ	DIAZ	MEDINA	TOTAL
33.	January	5416	7826	8542	3846	
34.	February	3829	27174	8026	3527	
35.	March	5522	7649	8634	14022	
36.	April	8312	8640	8426	4124	
37.	May	5082	7321	7947	3769	
38.	June	12578	13760	11230	5137	
39.	July	3984	7024	7640	3512	
40.	August	13226	3654	7237	3026	
41.	September	4759	6925	17926	3048	
42.	October	5726	7940	8747	3912	
43.	November	5247	8125	8837	4046	
44.	December	14275	16847	15522	7529	
	TOTAL	45.	46.	47.	48.	49.

18 © Copyright South-Western Publishing Co.

LESSON 10 — Subtraction Facts

Addition is the first fundamental operation in mathematics. Subtraction, the process of finding the difference between two numbers, is the second fundamental operation. The top number is the *minuend*. The bottom number is the *subtrahend*. The answer is the *difference* or *remainder*. These terms are illustrated in the example at the left.

This exercise contains 100 basic combinations for subtraction. *Study them, left to right, up and down, and on the diagonal*, to learn them thoroughly.

```
14  minuend
 5  subtrahend
 9  difference
```

EXERCISE 10 / Basic Time — $\frac{3}{4}$ Minute

Estimated time to obtain a basic score of 100.

Subtract: (2 each)

	a	b	c	d	e	f	g	h	i	j
1.	0/0	1/0	2/0	3/0	4/0	5/0	6/0	7/0	8/0	9/0
2.	1/1	2/1	3/1	4/1	5/1	6/1	7/1	8/1	9/1	10/1
3.	2/2	3/2	4/2	5/2	6/2	7/2	8/2	9/2	10/2	11/2
4.	3/3	4/3	5/3	6/3	7/3	8/3	9/3	10/3	11/3	12/3
5.	4/4	5/4	6/4	7/4	8/4	9/4	10/4	11/4	12/4	13/4
6.	5/5	6/5	7/5	8/5	9/5	10/5	11/5	12/5	13/5	14/5
7.	6/6	7/6	8/6	9/6	10/6	11/6	12/6	13/6	14/6	15/6
8.	7/7	8/7	9/7	10/7	11/7	12/7	13/7	14/7	15/7	16/7
9.	8/8	9/8	10/8	11/8	12/8	13/8	14/8	15/8	16/8	17/8
10.	9/9	10/9	11/9	12/9	13/9	14/9	15/9	16/9	17/9	18/9

© Copyright South-Western Publishing Co.

TEST 10

Name _____ Total Score _____
Date _____ Basic Score 100
Basic Time – $\frac{3}{4}$ Minute Improvement Score _____

Subtract: (2 each)

	a	b	c	d	e	f	g	h	i	j
1.	9 − 7	10 − 1	13 − 5	0 − 0	12 − 5	9 − 9	11 − 5	12 − 4	5 − 0	16 − 8
2.	6 − 0	9 − 0	10 − 2	13 − 8	11 − 2	14 − 5	8 − 3	6 − 5	16 − 7	13 − 7
3.	15 − 6	11 − 9	9 − 4	4 − 2	14 − 6	8 − 4	3 − 1	7 − 3	12 − 9	9 − 1
4.	14 − 8	8 − 7	1 − 0	5 − 5	7 − 6	5 − 2	12 − 6	2 − 2	4 − 3	8 − 6
5.	11 − 8	7 − 0	9 − 5	8 − 1	13 − 9	10 − 6	11 − 7	6 − 6	14 − 7	17 − 9
6.	9 − 8	11 − 3	16 − 9	5 − 3	7 − 7	12 − 3	7 − 2	17 − 8	10 − 8	13 − 4
7.	9 − 3	8 − 8	10 − 3	3 − 3	8 − 2	6 − 4	5 − 1	1 − 1	12 − 7	8 − 5
8.	4 − 1	6 − 1	7 − 5	8 − 0	10 − 9	10 − 4	18 − 9	7 − 4	4 − 4	6 − 3
9.	12 − 8	9 − 2	2 − 0	6 − 2	3 − 0	11 − 6	4 − 0	10 − 5	15 − 9	3 − 2
10.	11 − 4	2 − 1	14 − 9	15 − 7	10 − 7	5 − 4	13 − 6	15 − 8	7 − 1	9 − 6

LESSON 11: Checking Subtraction

Checking subtraction through addition is easy. This is done by adding the *difference* to the *subtrahend*. The sum of these two numbers must equal the *minuend* if the answer is correct.

You may need to "borrow" from the next column to the left. If 46 is subtracted from 81, as shown in the example at the right, the 6 cannot be subtracted from 1. Therefore a 10 is borrowed from the "tens" column and is added to the 1, making 11. Then 6 is subtracted from 11 to get 5. Four is subtracted from the remaining 7 in the "tens" column to get 3, and the answer is 35. The check of the answer is also shown in the example.

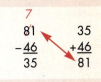

EXERCISE 11/Basic Time – 6 Minutes

Estimated time to obtain a basic score of 100.

Write the differences and check the answers: (202 points)

	a	b	c	d	e	f	g	h	
1.	91 45	124 74	102 57	143 79	203 98	117 67	129 58	120 83	(3)
2.	40 25	71 42	85 73	124 93	84 56	131 86	68 47	116 57	(3)
3.	103 84	86 29	73 37	88 69	94 47	132 85	91 53	55 26	(3)

	a	b	c	d	e	
4.	10,342 7,264	8,229 6,789	6,228 5,769	10,422 8,360	8,327 4,679	(6)
5.	11,254 5,698	8,129 7,643	5,762 1,915	10,215 9,487	9,215 3,746	(6)
6.	8,201 2,405	9,473 3,245	7,692 1,038	9,520 4,371	9,000 8,376	(6)

	a	b	c	d	
7.	81,245 53,627	746,210 123,798	151,627 92,748	1,007,618 308,247	(10)

TEST 11

Name _____ Total Score _____
Date _____ Basic Score 100
Basic Time–5 Minutes Improvement Score _____

1. Write in the differences and check the answers:

a	b	c	d	e	f	g	h	i	j
64	93	81	114	102	75	116	77	113	93
29	57	54	78	65	36	89	48	24	56

(5)

2. Stephen is 19 years old. His grandfather is 72 years old. His father is 43 years old.

   a. How many years younger is Stephen than his grandfather? 2a. _____ (10)

   b. How many years younger than his father? 2b. _____ (10)

   c. How many years older is his grandfather than his father? 2c. _____ (10)

3. Write in the differences and check the answers:

a	b	c	d	e	f
5,168	9,038	7,463	6,597	7,185	8,656
3,583	7,149	5,675	2,748	4,677	2,458

(10)

4. A market's June sales amounted to $647,894; July sales, $812,635; August sales, $756,946.

   a. What were the total sales for the three months? 4a. _____ (20)

   b. July sales were how much more than June sales? 4b. _____ (20)

   c. August sales were how much more than June sales? 4c. _____ (20)

22 © Copyright South-Western Publishing Co.

LESSON 12: Installment Buying Compared with Cash Buying

People who do not have ready cash often buy furniture and equipment for their homes on the installment plan. This plan provides for payment of the purchase price in small amounts over a period of months. Cash prices are almost always lower than installment prices; thus, you and your family can save money by purchasing with cash.

In this exercise, horizontal subtraction must be used to find the difference between the installment price and the cash price—the amount that is saved through a cash purchase. Record the saving (the difference) in the right-hand column. Be careful to subtract only those digits that are in the same position in each column of numbers.

EXERCISE 12/Basic Time—4 Minutes

Estimated time to obtain a basic score of 100.

Subtract horizontally: (10 each)

	ITEM	INSTALLMENT PRICE	CASH PRICE	SAVING
1.	Refrigerator	799 99	698 98	
2.	Television set	987 50	899 25	
3.	Vacuum cleaner	225 40	179 50	
4.	Washer	629 90	589 99	
5.	Microwave cart	89 00	74 95	
6.	Electric dryer	589 00	459 45	
7.	Student's desk	215 75	169 95	
8.	Living room chair	339 55	277 77	
9.	Sewing machine	393 72	325 39	
10.	Cassette player	239 95	199 50	
11.	Dinette	805 19	659 99	
12.	Rocking chair	160 89	138 70	
13.	Twin bed	189 00	149 98	
14.	Bedroom dresser	268 95	236 40	
15.	Electric cooktop unit	295 00	279 50	
16.	Game table	569 95	448 45	
17.	Metal bookcase	229 00	193 25	
18.	Bunk bed	324 88	282 50	
19.	Sofa	816 12	686 55	
20.	Camcorder	985 00	869 95	

Name _____
Date _____
Basic Time—4 Minutes

Total Score _____
Basic Score __100__
Improvement Score _____

Record the saving: (10 each)

	ITEM	INSTALLMENT PRICE		CASH PRICE		SAVING	
1.	Area rug	198	50	179	95		
2.	Dinnerware set	299	00	275	50		
3.	Mirror	85	00	79	95		
4.	Electric range	730	66	577	77		
5.	Air conditioner	953	98	799	99		
6.	Swivel chair	189	80	159	50		
7.	Home computer	859	25	689	99		
8.	AM-FM radio	67	63	57	80		
9.	Table lamp	88	15	76	65		
10.	Ceiling fan	102	50	93	49		
11.	Electric timer	48	99	39	95		
12.	Headboard	159	84	129	95		
13.	Computer desk	74	95	62	50		
14.	Recliner chair	477	95	419	25		
15.	Bedroom night stand	108	55	88	98		
16.	Bathroom scale	69	70	56	42		
17.	Loveseat	413	97	369	95		
18.	Portable stereo	287	07	242	25		
19.	Coffee maker	63	19	49	95		
20.	Cordless telephone	78	94	65	50		

© Copyright South-Western Publishing Co.

LESSON 13: Bank Deposits and Checks

To avoid overdrawing a checking account, you must keep a record of all deposits made and all checks issued. This is done in the register of the checkbook. When a deposit is made, the amount is entered in the register and added to the previous balance. Likewise, when a check is written, the amount is written in the register and subtracted from the previous balance. The last balance is the amount still on deposit.

In this exercise, the register has been abbreviated. The date, the name of the person to whom the check was issued, and the purpose of payment have been omitted. As one error will make the balances thereafter incorrect, strive for absolute accuracy. Be sure to subtract each check from the previous balance and to add each deposit.

EXERCISE 13/Basic Time—4 Minutes

Estimated time to obtain a basic score of 100.

Fill in the balances: (8 each)

Item	Amount		Item	Amount
Deposit	7,500.00		Balance for'd	
Check #1	387.40		Check #11	51.35
Balance	7,112.60		Balance	7,373.58
Check #2	45.37		Check #12	33.87
Balance	7,067.23		Balance	7,339.71
Check #3	125.75		Deposit	418.75
Balance	6,941.48		Balance	7,758.46
Deposit	375.43		Check #13	10.15
Balance	7,316.91		Balance	7,748.31
Check #4	18.70		Check #14	72.83
Balance	7,298.21		Balance	7,675.48
Check #5	12.50		Check #15	227.92
Balance	7,285.71		Balance	7,447.56
Check #6	53.48		Check #16	84.27
Balance	7,232.23		Balance	7,363.29
Deposit	234.29		Deposit	135.20
Balance	7,466.52		Balance	7,498.49
Check #7	217.65		Check #17	72.60
Balance	7,248.87		Balance	7,425.89
Check #8	91.34		Check #18	49.38
Balance	7,157.53		Balance	7,376.51
Check #9	13.60		Check #19	138.96
Balance	7,143.93		Balance	7,237.55
Deposit	418.50		Deposit	187.50
Balance	7,562.43		Balance	7,425.05
Check #10	137.50			
Balance forward	7,424.93			

TEST 13

Name _____ Total Score _____
Date _____ Basic Score 100
Basic Time–4 Minutes   Improvement Score _____

Fill in the balances: (8 each)

Deposit	1,250	00
Check #1	38	75
Balance	1,211	25
Check #2	76	58
Balance	1,134	67
Check #3	135	63
Balance	999	04
Deposit	441	60
Balance	1,440	64
Check #4	62	15
Balance	1,378	49
Check #5	68	70
Balance	1,309	79
Check #6	47	25
Balance	1,262	54
Check #7	358	98
Balance	903	56
Deposit	239	50
Balance	1,143	06
Check #8	85	75
Balance	1,057	31
Check #9	92	32
Balance	964	99
Check #10	140	68
Balance	824	31
Deposit	256	35
Balance Forward	1,080	66

Balance for'd	1,080	66
Check #11	73	15
Balance	1,007	51
Check #12	47	84
Balance	959	67
Deposit	283	45
Balance	1,243	12
Check #13	68	46
Balance	1,174	66
Check #14	235	15
Balance	939	51
Check #15	75	23
Balance	864	28
Check #16	24	25
Balance	840	03
Deposit	126	50
Balance	966	53
Check #17	58	00
Balance	908	53
Check #18	27	18
Balance	881	35
Check #19	59	27
Balance	822	08
Deposit	46	35
Balance	868	43

LESSON 14

Customers' Accounts (Accounts Receivable)

This lesson illustrates one more practical use for your ability to add and subtract accurately and speedily. Here is a group of customers' accounts, or *accounts receivable*. An *account* is a ruled form in which amounts are recorded in a systematic manner. To save space here, only the money columns are shown; date and explanation columns are omitted.

In the first column, headed DEBIT, charges made by the customer are written. In the second column, headed CREDIT, payments on account are recorded. After every entry, the new account balance is shown in the third column. A debit amount is added to the previous balance in a customer's account. A credit amount is subtracted from the previous balance.

EXERCISE 14/Basic Time–5 Minutes

Estimated time to obtain a basic score of 100.

Find the balance after each debit or credit: (5 each)

1. **Helen Herbert**

DEBIT	CREDIT	BALANCE
44.19		44.19
132.22		
	100.00	
50.49		
	44.19	
	32.22	

2. **Kathy Brightman**

DEBIT	CREDIT	BALANCE
204.49		204.49
	150.00	
36.67		
	54.49	
109.19		
	100.00	

3. **Hart & Cummings**

DEBIT	CREDIT	BALANCE
345.19		345.19
	124.35	
352.15		
	320.84	
	100.00	
67.97		

4. **Ana Guerra**

DEBIT	CREDIT	BALANCE
78.09		78.09
	50.00	
96.76		
	3.48	
	28.09	
39.72		

5. **William F. McCoy**

DEBIT	CREDIT	BALANCE
135.05		135.05
	27.50	
	100.00	
19.95		
21.45		
36.59		
	50.00	
75.65		

6. **The American Supply Co.**

DEBIT	CREDIT	BALANCE
1,111.78		1,111.78
121.18		
	1,000.00	
50.43		

7. **Lois Stein**

DEBIT	CREDIT	BALANCE
24.72		24.72
	4.50	
	20.22	
92.41		

8. **Brown-Seely Corporation**

DEBIT	CREDIT	BALANCE
65.05		65.05
116.56		
92.41		
	65.05	
43.08		
	116.56	
30.44		
	250.00	

Name		Total Score	
Date		Basic Score	100
Basic Time–5 Minutes		Improvement Score	

Find the new balance after each debit or credit entry. Remember that in a customer's account a debit is added to and a credit is subtracted from the previous balance. (5 each)

1. **Ruby North**

DEBIT	CREDIT	BALANCE
34 09		34 09
	4 04	30 05
	30 05	0 00
66 19		66 19
39 47		105 66

2. **Marie Perez**

DEBIT	CREDIT	BALANCE
44 06		44 06
100 67		144 73
	44 06	100 67
	100 67	0 00
30 34		30 34

3. **Peterson Brothers**

DEBIT	CREDIT	BALANCE
34 11		34 11
	7 17	26 94
62 91		89 85
	26 94	62 91
132 22		195 13
64 71		259 84

4. **Lawrence & Roche**

DEBIT	CREDIT	BALANCE
38 21		38 21
	5 20	33 01
264 44		297 45
	133 01	164 44
401 72		566 16
	64 44	501 72
	400 00	101 72
129 09		230 81

5. **Ginger Spangler**

DEBIT	CREDIT	BALANCE
67 72		67 72
304 11		371 83
	150 00	221 83
	75 00	146 83
88 94		235 77

6. **J. F. McArthy & Sons**

DEBIT	CREDIT	BALANCE
78 19		78 19
	2 27	75 92
66 72		142 64
34 01		176 65
	100 00	76 65

7. **H. G. Fischer & Company**

DEBIT	CREDIT	BALANCE
107 67		107 67
	100 00	7 67
44 04		51 71
132 96		184 67
	57 67	127 00
	50 00	77 00

8. **Debra Sidlowski**

DEBIT	CREDIT	BALANCE
138 44		138 44
	100 00	38 44
62 11		100 55
46 51		147 06
	38 44	108 62
107 72		216 34
29 89		246 23
	150 00	96 23

LESSON 15 Application Problems

Both addition and subtraction are used in solving the problems in this exercise. Before starting to solve a word problem, carefully decide which operation is to be used.

Example A Mr. Lerner earns $26,400 a year while Mrs. Lerner earns $32,600 a year. What is their total income?
$26,400 + $32,600 = $59,000

Example B How much more does Mrs. Lerner earn than Mr. Lerner?
$32,600 − $26,400 = $6,200 more

EXERCISE 15/Basic Time−7 Minutes
Estimated time to obtain a basic score of 100.

Solve these problems. Write the answers in the spaces on the right. (20 each)

1. An automobile salesman will allow Tom Gartner a trade-in allowance of $3,865 toward the purchase of a used car priced at $10,749. How much difference will Tom have to pay in order to buy the car?   1. _____

2. In West City there were two big games on the same day. At the football game there were 47,117 people; at the baseball game, 62,122. How many more people attended the baseball game than attended the football game?   2. _____

3. A hiker carried a tent weighing 9 pounds, a pack weighing 38 pounds, and a sleeping bag weighing 4 pounds. How much weight did the hiker carry?   3. _____

4. On January 1 Alice Mukuda's automobile mileage reading was 10,846. On December 31 the reading was 39,367. How many miles was Alice's automobile driven during the year?   4. _____

5. An automobile dealer sold 7 cars on Monday, 6 on Tuesday, 8 on Wednesday, 13 on Thursday, 24 on Friday, and 9 on Saturday. How many automobiles were sold during these six days?   5. _____

6. At the beginning of the week, a gasoline pump showed a reading of 27,379. At the end of the week, the gasoline pump meter showed 40,887. According to the meter, how many gallons of gasoline were sold through this pump during the week?   6. _____

7. In the first of three offices, there were 32 desks; in the second, 26; and in the third, 41. How many desks were in the three offices?   7. _____

8. The population of Kenton is 13,587 and of Corinth 70,137. By how much is the population of Corinth greater than that of Kenton?   8. _____

9. How many gallons of gasoline can be refined from a 42-gallon barrel of crude oil which also yields the following: 20 gallons of fuel oil, 4 gallons of kerosene, 2 gallons of lubricating oil, 3 gallons of miscellaneous products, and 2 gallons of waste?   9. _____

10. A 17-foot canoe will support 750 pounds without sinking. If it were occupied by three persons weighing 130 pounds, 203 pounds, and 196 pounds, how much more weight should it hold without sinking?   10. _____

TEST 15

Name _____ Total Score _____

Date _____ Basic Score 100

Basic Time–7 Minutes Improvement Score _____

Solve these problems. Write the answers in the spaces on the right. (20 each)

1. A baker needed 626 pounds of materials to produce 548 pounds of bread. What is the difference between the total weight of the materials and the total weight of the bread?

   1._____

2. A nurse at City General Hospital counted 17 beds in one ward, 28 in another, and 39 in a third. How many beds did the nurse count in these wards?

   2._____

3. A camper was driven 247 miles the first day and 151 miles the second day. How many miles was it driven in these two days?

   3._____

4. If the mileage reading of an automobile was 26,940 at the beginning of the day and 27,279 at the end of the day, how many miles was the automobile driven that day?

   4._____

5. An-Ping Shen paid $63 for her automobile license, $3,580 income tax, and $1,795 property tax. What was her total expenditure for these items?

   5._____

6. The area of Alaska is 586,400 square miles. The area of Texas is 267,339 square miles. In square miles, how much larger is Alaska than Texas?

   6._____

7. The Jennings Manufacturing Company has liabilities totaling $697,592 and assets amounting to $904,325. By how much do the assets exceed the liabilities?

   7._____

8. During one week, a delivery truck was driven 42 miles on Monday, 78 miles on Tuesday, 36 miles on Wednesday, 91 miles on Thursday, 74 miles on Friday, and 37 miles on Saturday. How many miles was the truck driven that week?

   8._____

9. Graton Community College requires 62 semester units for graduation with an Associate Arts degree. If Steve Magosci has earned 16, 17, and 14 units during his three semesters at the college, how many units must he earn his fourth semester in order to graduate?

   9._____

10. At the beginning of November, the balance in a checking account was $784. During the month, checks for $39, $176, $18, $247, and $52 were written. Find the balance after these checks were deducted.

    10._____

© Copyright South-Western Publishing Co.

LESSON 16
Multiplication, 196 Key Combinations

Multiplication is the third basic operation in mathematics. It is a shortcut method of addition. For example, if you correctly add four 3's, the result will be 12—the same as 3 multiplied by 4. As in addition, reverse-checking is a good method of ensuring accuracy: 4 × 3 = 12 and 3 × 4 = 12. The top (or first) number is the *multiplicand;* the bottom (or second) number is the *multiplier;* the result of multiplication is the *product*. See the example at the right. Learn thoroughly the multiplication facts from 2 through 15.

In the form below, there are spaces for the products of 196 combinations in multiplication. Write the products in the spaces. Each square is to be filled in with the product that is the result of multiplying the number in the top row by the number in the left column.

```
  3
  3    3 multiplicand
  3   x4 multiplier
 +3   12 product
 12
```

EXERCISE 16/Basic Time – 1 Minute
Estimated time to obtain a basic score of 100.

Write the products in the squares as indicated: (1 each, 196 points)

	×	a 2	b 3	c 4	d 5	e 6	f 7	g 8	h 9	i 10	j 11	k 12	l 13	m 14	n 15
1.	2														
2.	3														
3.	4														
4.	5														
5.	6														
6.	7														
7.	8														
8.	9														
9.	10														
10.	11														
11.	12														
12.	13														
13.	14														
14.	15														

TEST 16

Name
Date
Basic Time–3 Minutes

Total Score _____
Basic Score __100__
Improvement Score _____

Write products in squares as in the preceding exercise: (2 each)

	×	a 6	b 15	c 11	d 7	e 3	f 9	g 12	h 8	i 4	j 5
1.	4										
2.	3										
3.	9										
4.	12										
5.	8										
6.	11										
7.	6										
8.	15										
9.	7										
10.	5										

© Copyright South-Western Publishing Co.

LESSON 17: Multiplication with One-Digit Multipliers

Multiplying by 2, 3, 4, and 5 is often easier than multiplying by 6, 7, 8, and 9. If the multiplication facts have been thoroughly memorized, however, even these latter digits will present little, if any, difficulty.

At times you must carry numerals. You may write down the carry numerals as shown in the example at the right, or simply remember them. Use the method that works better for you. In the example, the carry numeral is 5.

```
  5
 28     (2 tens + 8 ones)
X 7  (      X 7       )
196    (14 tens + 56 ones)
```

Solution. (1) Multiply 8 (ones) by 7 to obtain 56 (ones). Write 6 below the 7 in the ones place and carry the 5 (tens).

(2) Multiply 2 (tens) by 7 to obtain 14 (tens). To the 14 (tens), add the carried 5 (tens) to obtain 19 (tens). Write 9 in the tens place and 1 in the hundreds place. The product is 196.

EXERCISE 17/Basic Time—8 Minutes
Estimated time to obtain a basic score of 100.

Multiply: (2 each)

	a	b	c	d	e	f	g	h	i	j
1.	56 × 2	25 × 3	37 × 4	22 × 5	54 × 2	43 × 3	56 × 4	62 × 5	65 × 4	97 × 3
2.	48 × 2	34 × 3	35 × 2	47 × 2	52 × 6	59 × 2	41 × 7	55 × 2	65 × 8	74 × 3
3.	73 × 8	64 × 8	42 × 5	86 × 3	45 × 7	52 × 8	59 × 4	84 × 6	58 × 4	69 × 3
4.	56 × 5	65 × 4	64 × 9	66 × 7	74 × 8	72 × 6	77 × 4	67 × 7	86 × 8	28 × 6
5.	87 × 5	84 × 7	88 × 8	94 × 8	99 × 6	98 × 9	78 × 5	86 × 6	93 × 8	96 × 7
6.	90 × 8	92 × 9	94 × 5	87 × 6	99 × 7	95 × 8	82 × 8	32 × 6	79 × 7	50 × 9
7.	52 × 9	83 × 6	94 × 7	87 × 7	65 × 9	52 × 7	35 × 9	93 × 7	88 × 7	76 × 8
8.	93 × 6	86 × 7	95 × 9	75 × 7	96 × 8	48 × 7	96 × 6	54 × 8	87 × 7	60 × 9
9.	95 × 7	96 × 9	98 × 8	97 × 8	79 × 9	91 × 8	88 × 7	92 × 6	89 × 8	87 × 9
10.	79 × 8	85 × 8	93 × 9	99 × 9	97 × 9	98 × 6	87 × 8	78 × 7	68 × 8	94 × 9

© Copyright South-Western Publishing Co.

TEST 17

Name _____ Total Score _____
Date _____ Basic Score __100__
Basic Time–7 Minutes Improvement Score _____

Multiply: (2 each)

	a	b	c	d	e	f	g	h	i	j
1.	52 × 3	82 × 7	67 × 2	32 × 8	96 × 4	42 × 8	72 × 9	91 × 6	96 × 5	35 × 9
2.	63 × 6	98 × 4	93 × 8	43 × 9	83 × 7	47 × 5	95 × 4	45 × 7	44 × 9	47 × 6
3.	53 × 9	56 × 8	58 × 7	60 × 5	59 × 9	67 × 7	65 × 6	67 × 6	69 × 7	62 × 9
4.	78 × 8	79 × 6	84 × 6	77 × 8	78 × 7	84 × 8	89 × 7	86 × 5	88 × 9	87 × 8
5.	94 × 7	98 × 7	99 × 6	95 × 7	98 × 6	95 × 8	94 × 6	93 × 9	90 × 8	96 × 9
6.	22 × 9	35 × 7	97 × 6	99 × 7	84 × 9	96 × 8	63 × 9	87 × 8	66 × 7	69 × 8
7.	88 × 7	42 × 8	86 × 7	84 × 8	93 × 8	70 × 8	78 × 8	75 × 7	85 × 9	86 × 9
8.	73 × 7	75 × 8	74 × 9	97 × 7	82 × 9	87 × 6	85 × 8	88 × 8	90 × 6	98 × 8
9.	93 × 6	97 × 8	96 × 9	98 × 9	98 × 7	95 × 9	83 × 6	86 × 8	89 × 7	99 × 8
10.	99 × 5	98 × 5	87 × 9	75 × 6	69 × 9	79 × 8	96 × 6	98 × 9	79 × 7	98 × 6

LESSON 18: Multiplication with Two- and Three-Digit Multipliers

When multiplying by a number that contains two or more digits, you must be careful to align the partial products properly. The right-hand digit of each partial product is written directly below the multiplying digit. This causes the partial products to be properly aligned so that tens will be added to tens, hundreds to hundreds, etc. The partial products are added together to obtain the final product, as shown in the example below.

Zeros
```
    5,746    multiplicand
  ×   312    multiplier
   11 492    first partial product
   57 46     second partial product
 1 723 8     third partial product
 1,792,752   product
```

EXERCISE 18/Basic Time—15 Minute
Estimated time to obtain a basic score of 100.

Multiply and insert commas in the product where applicable:

	a	b	c	d	e	f	g	h
1.	34 × 22	37 × 12	42 × 29	54 × 32	45 × 27	43 × 22	45 × 12	46 × 28 (4)
2.	95 × 29	69 × 26	98 × 25	89 × 23	81 × 27	99 × 38	98 × 37	95 × 24 (4)
3.	82 × 47	79 × 43	59 × 48	87 × 42	86 × 55	67 × 24	74 × 39	97 × 84 (4)

	a	b	c	d	e	f
4.	798 × 54	938 × 64	786 × 35	587 × 57	241 × 34	351 × 78 (6)
5.	651 × 88	597 × 67	489 × 36	792 × 91	563 × 99	938 × 45 (6)

	a	b	c	d
6.	7,864 × 381	8,955 × 279	8,926 × 698	6,847 × 457 (8)

Name _____ Total Score _____

Date _____ Basic Score 100

Basic Time–12 Minutes   Improvement Score _____

Multiply and insert commas in the product where applicable:

	a	b	c	d	e	f	g	h	
1.	63 32	46 23	95 21	42 31	73 34	93 43	85 12	94 27	(4)
2.	54 14	93 62	53 26	99 28	92 62	83 39	69 13	86 25	(4)
3.	96 38	86 79	89 88	93 29	84 35	78 46	98 85	97 57	(4)

	a	b	c	d	e	f	
4.	573 34	388 54	676 96	826 85	976 63	778 77	(6)
5.	854 37	957 84	575 53	597 89	978 65	978 99	(6)

	a	b	c	d	
6.	3,914 573	9,833 412	7,256 769	9,342 684	(8)

36 © Copyright South-Western Publishing Co.

LESSON 19
Multiplication with 0 in Multiplicand and in Multiplier

often cause errors in multiplication. Some people make errors when 0 is in a multiplication problem because (1) they forget that 0 multiplied by any number always equals 0, and (2) they carelessly misplace numerals in partial products.

For example, 4 × 0 simply means 0 + 0 + 0 + 0. This equals 0. Reversing the numerals in the problem gives 0 × 4, which means "not any" 4's. Zero is merely the symbol used to represent none. You can avoid having difficulty with 0's in multiplication if you will remember (1) zero means "not any," and (2) the right-hand digit of each partial product is placed directly below its multiplying digit.

The example below shows two ways of multiplying by zero. The second method is shorter and more commonly used.

Numbers ending in zero present the greatest

```
   3,405
  x2,001                    3,405
   3 405                   x2,001
     00 00         or       3 405
    000 0                   6 810 00
  6 810                     6,813,405
  6,813,405
```

EXERCISE 19/Basic Time—8 Minutes
Estimated time to obtain a basic score of 100.

Multiply and insert commas in the product:

	a	b	c	d	e
1.	3,002 9	5,006 8	4,007 7	6,008 6	7,009 5 (4)
2.	304 13	502 32	605 45	309 67	708 89 (6)
3.	4,003 87	5,006 79	7,005 98	8,090 65	9,070 86 (8)
4.	712 305	322 406	793 509	754 608	919 807 (10)
5.	6,400 6,005	7,500 8,090	4,030 7,040	7,809 9,060	8,069 5,004 (12)

© Copyright South-Western Publishing Co.

TEST 19

Name _____ Total Score _____

Date _____ Basic Score 100

Basic Time—10 Minutes Improvement Score _____

Multiply and insert commas in the product:

	a	b	c	d	e
1.	3,010 × 5	4,006 × 6	5,090 × 7	6,100 × 8	7,055 × 9 (4)
2.	907 × 28	908 × 37	804 × 46	806 × 57	604 × 79 (6)
3.	3,002 × 59	5,008 × 68	4,007 × 75	9,050 × 86	8,060 × 94 (8)
4.	688 × 707	775 × 609	632 × 508	849 × 407	609 × 509 (10)
5.	6,500 × 7,060	9,507 × 7,005	8,070 × 8,040	8,700 × 9,600	7,084 × 8,900 (12)

38 © Copyright South-Western Publishing Co.

LESSON 20: Multiplication by 10 and by Multiples of 10

Numbers ending in zero present the greatest possibilities for shortcuts in multiplication. Study the shortcuts listed below until you know them thoroughly.

To multiply any number by 10, simply attach a 0 to the right of the number.

To multiply any number by 20, multiply the number by 2 and attach a 0 to the right of the result.

To multiply any number by 30, multiply the number by 3 and attach a 0 to the right of the result.

Mentally complete the above set of rules through 40, 50, 60, 70, 80 and 90.

To multiply any number by 100, attach two 0's to the right of the number.

To multiply any number by 200, multiply the number by 2 and attach two 0's to the right of the result. See the example at the right.

```
    63
   200
12,600
```

Mentally complete the above set of rules through 900.

To multiply any number by 1,000, attach three 0's to the right of the number.

To multiply any number by 2,000, multiply the number by 2 and attach three 0's to the right of the result.

Mentally complete the above set of rules through 9,000.

EXERCISE 1/Basic Time – 3 Minutes
Estimated time to obtain a basic score of 100.

Multiply and insert commas in the product where applicable: (3 each for Nos. 1-20; 5 each for 21-22.)

	a	b	c
1.	63 x 10 =	51 x 7,000 =	135 x 10 =
2.	63 x 100 =	53 x 8,000 =	254 x 100 =
3.	63 x 1,000 =	57 x 9,000 =	362 x 1,000 =
4.	63 x 30 =	59 x 10 =	451 x 20 =
5.	63 x 40 =	61 x 20 =	532 x 30 =
6.	63 x 50 =	63 x 30 =	617 x 40 =
7.	63 x 60 =	65 x 40 =	781 x 50 =
8.	63 x 70 =	67 x 50 =	893 x 60 =
9.	63 x 80 =	69 x 60 =	918 x 70 =
10.	63 x 90 =	71 x 70 =	122 x 80 =
11.	63 x 200 =	73 x 80 =	266 x 90 =
12.	63 x 300 =	75 x 90 =	345 x 200 =
13.	63 x 400 =	76 x 100 =	456 x 300 =
14.	63 x 500 =	77 x 200 =	576 x 400 =
15.	63 x 600 =	79 x 300 =	677 x 500 =
16.	63 x 700 =	81 x 400 =	789 x 600 =
17.	63 x 800 =	83 x 500 =	803 x 700 =
18.	63 x 900 =	85 x 600 =	904 x 800 =
19.	98 x 2,000 =	87 x 700 =	904 x 900 =
20.	98 x 3,000 =	89 x 800 =	102 x 2,000 =

	a.	b.
21.	124 x 4,000 =	137 x 5,000 =
22.	242 x 6,000 =	999 x 7,000 =

TEST 20

Name _____
Date _____
Basic Time–2½ Minutes

Total Score _____
Basic Score __100__
Improvement Score _____

Multiply and insert commas in the product where applicable: (3 each for Nos. 1-20; 5 for 21-22.)

	a	b	c
1.	16×10=	22×10=	195×200=
2.	61×10=	32×10=	185×300=
3.	27×100=	42×20=	289×800=
4.	71×100=	52×30=	278×600=
5.	38×1,000=	62×30=	884×700=
6.	81×1,000=	72×40=	884×900=
7.	49×20=	72×50=	348×800=
8.	91×200=	82×60=	377×700=
9.	21×300=	82×70=	677×800=
10.	32×400=	83×80=	553×300=
11.	53×500=	84×90=	456×400=
12.	34×600=	94×100=	465×400=
13.	43×700=	94×1,000=	564×500=
14.	65×800=	49×20=	546×600=
15.	54×900=	94×30=	674×700=
16.	76×2,000=	59×90=	687×900=
17.	65×3,000=	98×80=	781×800=
18.	87×4,000=	77×10=	780×800=
19.	76×5,000=	67×1,000=	345×800=
20.	98×6,000=	89×40=	354×200=

	a	b
21.	561×5,000=	824×6,000=
22.	942×7,000=	924×7,000=

LESSON 21

Application Problems

These word problems require the use of addition, subtraction, or multiplication. Carefully decide which of these to use before solving each problem below.

Example A college student reads on the average 300 pages a week. If a term lasts 15 weeks, how many pages does this student read in one term?
300 x 15 = 4,500 pages

EXERCISE 21/Basic Time–8 Minutes
Estimated time to obtain a basic score of 100.

Solve these problems. Write the answers in the spaces on the right. (20 each)

1. The deputy sheriff had the following scores in target practice: 9, 7, 8, 9, 6, 10, 7, 9. What was the total score?

    1._____

2. The bituminous coal from Franklin County, Illinois, has an average heat value (British Thermal Units) per pound of 11,930. Lignite coal from Houston County, Texas, has an average of 7,140 BTU's per pound. This is a difference of how many BTU's per pound?

    2._____

3. How much weight is handled by a person who moves 765 boxes of apples if each box weighs 60 pounds?

    3._____

4. Two herds of cattle were counted. The first contained 47 steers; the second, 56. How many steers were in these two herds?

    4._____

5. At the beginning of the year, there were 68,809 people registered at the county unemployment office and 53,910 registered at the end of the year. This is a decrease of how many?

    5._____

6. If a bakery sold an average of 800 cakes a day, how many cakes would it sell in a month in which it was open 27 days?

    6._____

7. If a person can type at an average speed of 64 words a minute, how many words can he or she type in 42 minutes?

    7._____

8. How far can a truck travel on 98 gallons of gasoline if it averages 13 miles to a gallon?

    8._____

9. The average attendance at the Strand Theater was 753 people for 468 consecutive performances. How many people attended during this period of time?

    9._____

10. In a new subdivision of Eaglespoint, there are 86 condominiums worth $239,725 each. What is the total value of these condominiums?

    10._____

Name _____ Total Score _____

Date _____ Basic Score __100__

Basic Time–7 Minutes Improvement Score _____

Solve these problems. Write the answers in the spaces on the right. (20 each)

1. The Wyoming Hikers Club walked 28 miles on Thursday, 27 miles on Friday, and 26 miles on Saturday. How many total miles did the club hike on Friday and Saturday?

   1. _____

2. A small airplane flies at an average ground speed of 158 miles an hour. How far will it travel in 17 hours?

   2. _____

3. Figure the difference if the subtrahend is 1,058 and the minuend is 1,951?

   3. _____

4. Last year Argon Equipment Company bought 63 new machines at $13,798 each. What was the total cost?

   4. _____

5. The demand for electric motors caused Universal Motors to deplete its stock and to ship the motors as fast as they were produced. During this time the company produced an average of 1,867 motors every day for 34 days. What is the total number of motors shipped during this period?

   5. _____

6. The Alert Answering Service handled 639 messages a day during February (28 days). What was the total number of messages handled for this month?

   6. _____

7. One section of a road cost three million dollars a mile for 92 miles. What was the total cost of constructing this section?

   7. _____

8. For the spring semester, 2,314 freshmen and 1,740 sophomores were enrolled at Denton University. What was the total enrollment for these two classes?

   8. _____

9. A new office center contains 19 offices of 900 square feet each, 25 offices of 1,200 square feet each, and 9 offices of 1,500 square feet each. This center contains how many square feet of office space?

   9. _____

10. A building being constructed is to have 376 windows. Each window is to contain 18 panes of glass. How many panes of glass will be needed?

    10. _____

© Copyright South-Western Publishing Co.

LESSON 22: Division Facts

Division is the fourth and last basic operation in mathematics. It is the process of finding how many times one number can be contained in another. In effect, division is the inverse (opposite) of multiplication. The number to be divided is the *dividend;* the number that does the dividing is the *divisor;* the result obtained by division is the *quotient*. To prove the quotient, multiply the divisor by the quotient to get the dividend.

See the example at the right. If 6 is divided by 2, the quotient is 3. Then to check the answer, multiply the divisor 2 by the quotient 3 to get the dividend 6.

```
        3 quotient
divisor 2)6 dividend
```

This exercise contains the 90 basic division facts arranged in random order. Since division is closely related to multiplication, a review of the multiplication tables will prove helpful. Before writing the quotients, do them mentally until you can complete them accurately in one minute. Then time yourself as you write the quotients. Test 22 may be used as a speed test.

EXERCISE 22/Basic Time—1 Minute

Estimated time to obtain a basic score of 100.

Divide: (2 each, 180 points)

	a	b	c	d	e	f	g	h	i
1.	4)0	5)5	3)9	1)4	1)5	2)4	9)9	6)36	1)3
2.	3)6	2)16	9)27	8)0	2)6	6)12	6)48	4)24	8)40
3.	7)42	9)45	5)10	1)6	8)64	4)32	6)54	6)6	2)0
4.	4)16	8)8	1)0	5)30	5)45	8)16	9)81	3)18	8)72
5.	2)10	7)21	7)7	9)63	8)56	9)36	8)32	4)36	9)0
6.	3)24	4)8	1)2	5)0	9)18	2)14	7)56	2)18	4)20
7.	5)20	7)35	6)0	1)9	7)63	8)48	6)30	3)27	5)40
8.	4)12	9)72	3)3	1)7	8)24	6)18	2)12	3)0	7)28
9.	5)15	4)28	7)14	5)25	3)21	9)54	1)8	3)15	6)42
10.	5)35	3)12	7)0	7)49	4)4	6)24	2)8	2)2	1)1

TEST 22

Name _____ Total Score _____
Date _____ Basic Score __90__
Basic Time–1 Minute   Improvement Score _____

Divide: (2 each, 180 points)

	a	b	c	d	e	f	g	h	i
1.	1)5̄	8)8̄	7)0̄	9)9̄	4)0̄	1)2̄	3)9̄	2)6̄	4)4̄
2.	2)8̄	5)1̄0̄	6)2̄4̄	7)6̄3̄	2)1̄0̄	4)1̄2̄	1)9̄	3)2̄7̄	1)1̄
3.	3)2̄4̄	4)1̄6̄	8)7̄2̄	8)3̄2̄	2)1̄2̄	9)4̄5̄	7)7̄	1)7̄	6)6̄
4.	7)2̄1̄	6)4̄2̄	9)0̄	9)3̄6̄	5)3̄5̄	7)5̄6̄	5)2̄5̄	5)1̄5̄	3)2̄1̄
5.	7)4̄9̄	5)2̄0̄	5)4̄5̄	7)2̄8̄	2)0̄	8)0̄	9)2̄7̄	8)4̄8̄	6)1̄2̄
6.	4)3̄2̄	5)4̄0̄	3)1̄5̄	2)1̄4̄	7)1̄4̄	1)6̄	2)1̄6̄	9)5̄4̄	2)1̄8̄
7.	8)2̄4̄	4)2̄8̄	3)1̄2̄	6)4̄8̄	1)3̄	7)3̄5̄	8)4̄0̄	6)3̄0̄	4)3̄6̄
8.	8)5̄6̄	5)5̄	4)2̄0̄	6)1̄8̄	3)1̄8̄	9)8̄1̄	8)6̄4̄	9)6̄3̄	9)1̄8̄
9.	8)1̄6̄	7)4̄2̄	4)2̄4̄	6)5̄4̄	9)7̄2̄	6)3̄6̄	5)3̄0̄	5)0̄	3)0̄
10.	2)4̄	1)4̄	4)8̄	1)8̄	2)2̄	1)0̄	3)3̄	6)0̄	3)6̄

LESSON 23
Division with Small Divisors

This exercise is limited to division by numbers up to and including 15. The method to be used here is known as *short division*. Only the quotients are to be written.

Start dividing on the left. Write each "carry-over" digit as a small numeral to the left of the next digit. Show any remainder in parentheses to the right of the quotient. See the example at the right.

Solution. As 10 contains two 5's, write 2 over the 0. The 8 contains one 5 with 3 "left over," so write 1 over the 8, and carry 3 to the left of the 9. As 39 contains 7 fives with 4 remaining, write 7 over the 9. Show the remainder, 4, in parentheses next to the quotient.

$$217(4)$$
$$5\overline{)1{,}08\,9}$$

Following is a popular method of checking accuracy in division. Notice that the dividend shown below (1,089) does equal the dividend in the example above.

Quotient × Divisor + Remainder = Dividend
217 × 5 + 4 = 1,089

EXERCISE 23/Basic Time–4 Minutes
Estimated time to obtain a basic score of 100.

Divide and show any remainder in parentheses next to the quotient: (198 points)

	a	b	c	
1.	2)8,423	3)6,373	4)8,497	(6)
2.	4)84,536	5)55,673	6)97,268	(6)
3.	7)77,950	8)69,208	9)11,106	(6)
4.	4)1,764,911	5)547,691	6)742,351	(6)
5.	5)173,855	6)284,156	7)932,617	(6)
6.	7)6,382,523	8)1,175,140	9)7,369,792	(10)
7.	10)154,110	11)57,388	12)25,816	(12)
8.	13)427,623	14)782,386	15)736,254	(14)

TEST 23

Name _____ Total Score _____
Date _____ Basic Score _99_
Basic Time—4 Minutes Improvement Score _____

Divide and show any remainder in parentheses next to the quotient: (198 points)

        a   b   c

1. 2)6,742 3)9,427 4)1,478 (6)

2. 5)65,675 6)27,462 7)69,378 (6)

3. 6)26,948 8)36,278 9)19,314 (6)

4. 5)124,692 7)678,449 8)345,671 (6)

5. 6)673,272 7)114,431 8)753,834 (6)

6. 7)1,823,976 8)2,545,138 9)1,957,295 (10)

7. 10)261,180 11)129,096 12)145,729 (12)

8. 13)211,926 14)134,554 15)146,798 (14)

LESSON 24 Long Division

Long division is the name applied to division by numbers that cannot readily be used as simple divisors. The individual products and remainders, because of their length, are written down.

In the example at the right, 159,872 (the *dividend*) is divided by 364 (the *divisor*) to give a *quotient* of 439 and a *remainder* of 76.

Solution. First, 364 goes into 1598 four times with a remainder. Write 4 above the 8 of the dividend. Then multiply this 4 by the divisor of 364. Write your answer, 1456, under 1598, and subtract to find the remainder of 142.

Bring down the next digit (7) from the dividend to get 1427. Divide 1427 by 364 to get 3 plus a remainder. Write 3 above the 7 of the dividend, and multiply the 3 by the divisor of 364. The answer, 1092, is subtracted from 1427 to give 335.

Bring down the 2 from the dividend to get 3352. Divide 3352 by 364 to get 9 plus a remainder. Write 9 above the 2 of the dividend, and multiply by the divisor of 364. The answer, 3276, is subtracted from 3352 to give a remainder of 76.

```
          439
364) 159,872
     145 6
      14 27
      10 92
       3 352
       3 276
          76
```

EXERCISE 24/Basic Time–6 Minutes

Estimated time to obtain a basic score of 100.

Divide: (198 points)

	a	b	c	
1.	368)49,307	473)59,368	746)83,549	(18)
2.	785)90,294	845)123,925	639)101,623	(18)
3.	3,412)51,709	9,063)134,263	4,716)84,627	(18)
4.	387)12,463	695)11,576	439)12,590	(12)

	Name	Total Score
	Date	Basic Score 100
	Basic Time–5 Minutes	Improvement Score

Divide: (201 points)

	a	b	c	
1.	362)4,827	593)6,524	728)9,518	(13)
2.	928)10,693	805)16,275	627)8,276	(13)
3.	1,351)63,475	2,876)92,467	4,589)136,784	(17)
4.	529)72,673	856)163,724	7,563)2,037,865	(24)

LESSON 25: Zeros in the Quotient

While dividing, be careful when placing 0's in the quotient. Because 0 serves as a place holder, an omitted 0 causes the quotient to be wrong.

If the number to be divided is smaller than the divisor after a digit is brought down from the dividend, a 0 must be placed in the quotient above that dividend digit.

Example A. After 18 is divided into 55 three times with a remainder of 1 and the 2 is brought down, the 12 is smaller than the 18. Therefore, write 0 above the 2, bring down the 6, and divide 18 into 126.

A division problem is not complete until a digit is placed in the quotient directly above the last digit in the dividend.

Example A
```
        307                307
    18)5,526           18)5,526
       5 4                5 4
       12        or       126
       00                 126
       126
       126
```

Example B. As the remainder 15 is smaller than the divisor 23, write 0 in the quotient above the 5.

Example B
```
         40
     23)935
        92
        15
```

EXERCISE 25/Basic Time – 10 Minutes
Estimated time to obtain a basic score of 100.

Divide and show any remainder in parentheses:

	a	b	c	
1.	5)1,830	7)1,442	9)3,663	(10)
2.	8)8,664	9)16,309	6)8,406	(10)
3.	14)28,098	25)45,315	39)58,387	(14)
4.	45)67,635	61)128,466	74)266,457	(18)

	a	b	
5.	82)16,457,482	95)66,528,880	(22)

Name _____ Total Score _____

Date _____ Basic Score __100__

Basic Time–8 Minutes Improvement Score _____

Divide and show any remainder in parentheses:

	a	b	c	
1.	16)4,880	23)9,361	34)27,268	(16)

| 2. | 45)225,287 | 67)268,536 | 78)546,702 | (18) |

| 3. | 86)776,580 | 92)372,682 | 197)682,977 | (18) |

	a	b	
4.	203)8,181,606	374)14,228,786	(22)

LESSON 26: Finding Averages

This lesson provides practice in averaging numbers; that is, finding a number that is representative of all the numbers in a set.

To find the average of a set of numbers, add the numbers and then divide the sum by the number of numbers in the set.

Example. Find the average of these numbers: 28, 47, 63, 75

Solution.
1. Add the numbers.
2. Divide the sum by the number of numbers in the set.

Example

Step 1

```
  28
  47   (4 numbers
  63    in the set)
 +75
 ---
 213
```

Step 2

$$53\frac{1}{4}$$ average

```
4)213
  20
  ---
  13
  12
  ---
   1
```
(Place the remainder over the divisor as a fraction)

EXERCISE 26 / Basic Time – 8 Minutes
Estimated time to obtain a basic score of 100.

Find the average for each set of numbers. Show any remainder over the divisor as a fractional part.

1. 27, 38, 52, 69, 74   1. _____(8)

2. 129, 76, 257, 342, 638, 221   2. _____(22)

3. 543, 628, 947, 463, 696, 382, 543, 623   3. _____(20)

4. 228, 469, 133, 546, 1,020, 314, 847   4. _____(16)

5. 36, 73, 56, 85, 92, 43, 61, 28, 73, 57, 42, 38, 31   5. _____(20)

6. The number of customers on consecutive days were as follows:
   296, 305, 326, 321, 310, 342, 356, 341, 363
   Find the average number of customers per day.   6. _____(20)

7. The individual weights of a soccer team were:
   204, 196, 189, 184, 176, 167, 164, 163, 160, 151, 145
   What was the average weight?   7. _____(24)

8. Find the average of the following daily sales:
   $211,768; $196,725; $205,689; $225,983; $189,658; $254,879   8. _____(34)

9. Find the average daily school attendance:
   1,479; 1,498; 1,505; 1,512; 1,468   9. _____(14)

10. Find the average of the following attendance at basketball games:
    7,825; 2,649; 5,684; 6,392; 3,476; 4,368; 2,556; 5,275   10. _____(22)

© Copyright South-Western Publishing Co.

TEST 26

Name _____   Total Score _____
Date _____   Basic Score __100__
Basic Time—8 Minutes   Improvement Score _____

1. Compare these football teams as to average weights: (18 each average)

POSITION	TEAM 1 WEIGHT	TEAM 2 WEIGHT	TEAM 3 WEIGHT
C.	151	150	165
R. G.	175	171	170
L. G.	197	165	176
R. T.	217	171	182
L. T.	172	180	184
R. E.	171	184	153
L. E.	142	155	163
Q. B.	168	156	146
R. H.	165	153	168
L. H.	144	140	154
F. B.	155	180	162
Total			
Average			

2. Compare these years as to average monthly sales: (49 each average)

MONTH	LAST YEAR	THIS YEAR
January	$ 96,354	$ 98,847
February	125,721	135,768
March	105,349	116,853
April	113,762	123,918
May	95,926	96,025
June	103,451	112,736
July	94,673	89,529
August	132,426	148,285
September	115,392	121,728
October	95,236	105,845
November	97,512	112,756
December	136,457	143,568
Total		
Average		

3. Compare departments of a store as to average number of customers: (6 each average)

DAY	SHOES	JEWELRY	APPLIANCES
1st	326	314	318
2d	302	296	325
3d	333	340	321
4th	357	338	313
Total			
Average			

4. Compare daily attendance averages: (10 each average)

DAY	GRADE 1	GRADE 2	GRADE 3
Monday	926	945	913
Tuesday	948	956	928
Wednesday	1,012	1,059	989
Thursday	1,015	1,057	992
Friday	996	998	974
Total			
Average			

© Copyright South-Western Publishing Co.

LESSON 27
Application Problems

EXERCISE 27/Basic Time—10 Minutes
Estimated time to obtain a basic score of 100.

Solve these problems. Write the answers in the spaces on the right. (20 each)

1. If a worker saved $27 each week, how much would he/she save in 48 weeks?   1._____

2. Nancy Kessler traveled 1,296 miles on a train. The train averaged 54 miles an hour. How many hours were needed for this train trip?   2._____

3. A total of 168 printer ribbons were used in a department containing 24 printers. On the average, how many ribbons were used on each machine?   3._____

4. How many articles are contained in 12 cases of merchandise if each case contains 12 cartons and each carton contains 2 dozen articles?   4._____

5. A sales representative reported the following mileage for last week: Monday, 21 miles; Tuesday, 109 miles; Wednesday, 95 miles; Thursday, 23 miles; Friday, 92 miles. What is the average mileage per day?   5._____

6. How many gallons of fuel will be used by a 4-engine airplane on a flight of 6 hours if each engine uses 56 gallons of fuel an hour?   6._____

7. The teacher has 12 boxes of pencils with 25 pencils in each box. If these pencils were divided evenly among 50 students, how many pencils would each person get?   7._____

8. A machine can print reports at an average rate of 576 an hour. How many reports can the machine print in a 5-day week if it is operated 3 hours a day?   8._____

9. John Leone bought a new car that was covered by a guarantee that remains effective for the first 24,000 miles the car is driven or for two years, whichever occurs first. During the first 12 months, he drove the car 18,000 miles. If he continues to average the same mileage each month, for how many more months can he expect his guarantee to remain in effect?   9._____

10. A worker keyed 1,886 words in 41 minutes. How many words a minute did the worker average?   10._____

Name _____ Total Score _____

Date _____ Basic Score __100__

Basic Time–8 Minutes Improvement Score _____

Solve these problems. Write the answers in the spaces on the right. (20 each)

1. A total of 884,304 nails was used in the construction of a building. If there were 69 nails to a pound, how many pounds of nails were required?

   1._____

2. A typist had the following speed scores for a series of typewriting timings: 27, 28, 26, 31, 30, 32, 37, 33, and 35. What was the average score for this typist?

   2._____

3. The combined weights of Gail and Audrey total 258 pounds. (a) If Gail weighs 125 pounds, who weighs more—Gail or Audrey? (b) By how much?

   3a._____

   3b._____

4. How many bags, each containing 4 dozen pears, can be filled from 10 cases of pears if each case holds 30 dozen?

   4._____

5. Four heirs equally divided an estate of $520,000. Later a missing heir was located. How much must each of the original four heirs give to the fifth heir so that the estate will be equally divided among them?

   5._____

6. A used car dealer has 5 cars priced at $3,250 each; 9 cars at $2,349 each; 10 cars at $11,990 each; 8 cars at $9,990 each; 7 cars at $7,890 each; and 6 cars at $5,690 each. What is the total value of these cars according to the prices listed?

   6._____

7. On an automobile trip, the Jacob family had 1,209 miles to travel. The family drove 391 miles the first day but only 87 miles the second day because of a delay caused by engine trouble. The third day they drove 493 miles. How much farther did they have to drive?

   7._____

8. Bruce earns $2,150 a month. Jason earns $450 a week. In one year (52 weeks) at these salaries, how much more will Bruce earn than Jason?

   8._____

9. Elaine Strauss bought a new car 23 months ago. The automobile was covered by a guarantee that remains effective for the first 50,000 miles the car is driven or for five years, whichever occurs first. Elaine has now driven the car for 28,750 miles. For how many more months can she expect the guarantee to remain in effect?

   9._____

10. Sound travels at about 1,087 feet per second. In how many seconds will the report of an explosion be heard by someone who is 16,305 feet away from it?

    10._____

1 Cumulative Review

Name _____

Complete each statement by supplying the missing word(s).

1. In any number, the second digit from the right in any three-digit group occupies the __?__ place.

2. A __?__ is substituted for each digit that is omitted in a number that is rounded.

3. Numbers that are added together are properly called __?__.

4. The answer in a subtraction problem is called the __?__.

5. The smaller number in a subtraction problem is called the __?__.

6. The answer in a multiplication problem is called the __?__.

7. In a multiplication problem, the number by which one multiplies is properly called the __?__.

8. The answer in a division problem is properly called the __?__.

9. To help determine whether an answer is reasonable, compare it to an estimate, which can be obtained by using __?__ numbers.

10. In rounding to hundreds, 51 becomes __?__.

1. _____
2. _____
3. _____
4. _____
5. _____
6. _____
7. _____
8. _____
9. _____
10. _____

Write the following as numbers with commas inserted.

11. Four hundred thousand, five hundred twenty-three.

12. Fifty million, seventy.

13. One hundred two million, one hundred seventy thousand, four hundred seventy-six.

14. Eighty-seven billion, three hundred fifty million, nine hundred ten.

11. _____
12. _____
13. _____
14. _____

Round each of the following numbers to the place indicated.

15. 35,643,298 to the nearest hundred thousand.

16. 127,483,867 to nearest million.

17. 208,684,620,942 to nearest billion.

18. 347,295,024,526 to nearest ten million.

15. _____
16. _____
17. _____
18. _____

Estimate the answer to each of these addition problems. Show with a plus or minus sign if your estimate should be increased or decreased.

19.	370	20.	4,875	21.	23,348
	988		5,909		48,581
	653		867		8,105
	<u>66</u>		4,098		29,900
			<u>519</u>		4,834
					<u>43,976</u>

19. _____
20. _____
21. _____

© Copyright South-Western Publishing Co.

1 Cumulative Review

Estimate the answer to each of these subtraction problems. Show with a plus or minus sign if your estimate should be increased or decreased.

22. 7,949
    4,312

23. 69,466
    47,987

24. 748,960
     92,285

22. _____
23. _____
24. _____

Estimate the answer to each of these multiplication problems. Show with a plus or minus sign if your estimate should be increased or decreased.

25. 709
     82

26. 517
    435

27. 748,960
     1,017

25. _____
26. _____
27. _____

Estimate the answer to each of these division problems. Show with a plus or minus sign if your estimate should be increased or decreased.

28. 79)66,075

29. 65)591,440

30. 837)5,592,547

28. _____
29. _____
30. _____

Add. Watch for combinations that add to 10.

31. 53
    67
    42

32. 74
    32
    28

33. 28
    43
    82
    47
    63

34. 44
    57
    63
    52
    58

31. _____
32. _____
33. _____
34. _____

Add. Check each answer by the reverse-order check.

35. 8,619
    8,197
    5,906
    9,181
    3,915

36. 54,904
     5,591
    95,221
    92,813
     9,754

37. 89,724
    77,672
     4,728
    94,696
    97,234

35. _____
36. _____
37. _____

1 Cumulative Review

Name _____

Add horizontally and vertically.

38.	17	24	86	78	44	
39.	76	50	42	91	31	
40.	59	85	36	83	56	
41.	96	56	78	29	79	
42.	64	73	34	37	64	
43.	93	71	87	82	34	
	44.	45.	46.	47.	48.	49.

38. _____
39. _____
40. _____
41. _____
42. _____
43. _____
44. _____
45. _____
46. _____
47. _____
48. _____
49. _____

Subtract and check each answer.

50. 937,363
    627,682

51. 9,083,247
      732,373

52. 5,438,200
    3,549,945

50. _____
51. _____
52. _____

Show the balance after each check or deposit.

    Balance $450.37
    Deposit   962.91

53. Balance

    Check No. 401   109.00

54. Balance

    Check No. 402   31.45

55. Balance

    Check No. 403   273.00

56. Balance

    Deposit   850.05

57. Balance

    Check No. 404   291.35

58. Balance

    Check No. 405   325.15

59. Balance

53. _____
54. _____
55. _____
56. _____
57. _____
58. _____
59. _____

1 Cumulative Review

Subtract horizontally.

60. 531 − 283 =

61. 9,385 − 4,547 =

62. 4,304 − 1,868 =

63. 27,283 − 15,784 =

Multiply. Prove each answer.

64. 8507
     543

65. 8,931
     948

66. 8,721
     6,042

60. _____
61. _____
62. _____
63. _____

64. _____
65. _____
66. _____

Multiply. Use the shortcut method of appending zeros.

67. 756 × 80 =

68. 6,920 × 4,000 =

69. 1,093 × 500 =

70. 9,800 × 3,000 =

71. 8,234 × 9,000 =

72. 46,500 × 70,000 =

67. _____
68. _____
69. _____
70. _____
71. _____
72. _____

Divide. Check each answer by multiplying. Show any remainder in parentheses after the quotient.

73. 68)90,304

74. 94)85,643

73. _____
74. _____

The quarterly sales made by five salespeople for January, February, and March were: Dennis Alder, $87,000; Sheila Blakely, $94,500; Craig Cessna, $112,500, Maria Dietrich, $118,500, and James Eton, $114,000.

Use the above information to compute the following amounts.

75. Total quarterly sales (all salespeople)

76. Average quarterly sales (all salespeople)

77. Average monthly sales (all salespeople)

Average monthly sales (each salesperson):

78. Dennis Alder

79. Sheila Blakely

80. Craig Cessna

81. Maria Dietrich

82. James Eton

75. _____
76. _____
77. _____
78. _____
79. _____
80. _____
81. _____
82. _____

LESSON 28: Signed Numbers—Addition and Multiplication

The rules for adding *signed numbers* are as follows. (1) To add numbers with like signs, add the numerical values and attach the common sign to the total. (2) To add two numbers with unlike signs, find the difference in the numerical values and attach the sign of the number with the larger numerical value to the total. A number that does not have a sign is assumed to be a positive number.

Example A	Example B	Example C
−8	−43	−5
−5	+22	2
−6	−21	−8
−19		9
		−2

The rules for multiplying signed numbers are as follows. (1) To multiply two signed numbers, find the product of their numerical values. If the factors (the two numbers multiplied) have like signs, the product is positive. If the factors have unlike signs, the product is negative. (2) When multiplying more than two signed numbers, the product is positive if there is an even number of negative factors; the product is negative if there is an odd number of negative factors. Parentheses may be used instead of a times sign to show multiplication.

Example D
(−9)(−6) = +54

Example E
(−9)(6) = −54

Example F
(−6)(−4)(7)(−2) = −336

EXERCISE 28/Basic Time—9 Minutes

Estimated time to obtain a basic score of 100.

Add these signed numbers: (6 each) (Total: 204 points)

1. +12
   −5

2. +9
   +7

3. +14
   −25

4. −9
   +23

5. −15
   −32

6. +18
   −46
   +20

7. −16
   −40
   −31

8. −85
   32
   27

9. 26
   14
   −39
   −15

10. −8
    12
    −25
    18

11. −15
    −23
    1
    21

12. −16
    −33
    75
    −26

13. −9
    17
    26
    −35
    57

14. 56
    −67
    −44
    31

Multiply these signed numbers: (10 each)

15. +16
    +5

16. −34
    −9

17. −52
    6

18. 79
    −8

19. 68
    7

20. −45
    −5

21. 12
    −6

22. −37
    8

23. −64
    −7

24. −92
    −9

25. (−2)(5)(−3)(−4) =

26. (4)(−3)(5)(−6)(2) =

TEST 28

Name _____ Total Score _____
Date _____   Basic Score 100
Basic Time–9 Minutes   Improvement Score _____

Add these signed numbers: (6 each)

1. +2 4   2. –1 9   3. +4 6   4. –3 6   5. –8   6. –2 7
   +9   –8   –7   +3 7   –1 7   +4 4

7. –2 6   8. 2 8   9. 7 7   10. –2 6   11. 3 8
    5 8   –5 7   4 4   –5 4   –2 6
   –3 2   4 3   –3 9   2 8   4 7
                                                 –3 4

12. –7   13. 2 4   14. 2 5   15. –1 2   16. –4 7
     2 4   –3 2   –4 2   –9   5 3
     3 2   –1 0   6 6   1 7   –1
    –5 0   1   –1 7   –2 4   3 5
                                      4 8   –2 2

Multiply these signed numbers: (8 each)

17. +2 5   18. –4 7   19. +5 9   20. –1 8   21. +3 6
     +5   –8   –6   +7   –9

22. 8 2   23. –4 6   24. –9 3   25. 6 3   26. –8 1
      6   –5   –7   –9   8

27. (–5)(–3)(4)(–6) =

28. (–4)(6)(–3)(–5)(–2) =

29. (3)(–2)(5)(–2)(2)(–4)(5)(–3)(–2) =

60 © Copyright South-Western Publishing Co.

LESSON 29: Signed Numbers—Division and Subtraction

To divide signed numbers, find the quotient of their numerical values. If the numbers have like signs, the quotient is positive. If the numbers have unlike signs, the quotient is negative. Note that $\frac{10}{5}$ means the same as $5\overline{)10}$ or $10 \div 5$.

Example A
$\frac{10}{5} = 2$
Check: $5 \times 2 = 10$

Example B
$\frac{10}{-5} = -2$
Check: $-5 \times -2 = 10$

Example C
$\frac{-10}{5} = -2$
Check: $5 \times -2 = -10$

Example D
$\frac{-10}{-5} = 2$
Check: $-5 \times 2 = -10$

To subtract signed numbers, change the sign of the subtrahend and add. "Subtract +9 from +17" (Example E below) becomes "Add the signed numbers 17 and –9."

Example E
Subtract: +17
         +9
Solution: +17
         –9
Answer: + 8

Example F
Subtract: –17
         –9
Solution: –17
         +9
Answer: – 8

Example G
Subtract: +17
         –9
Solution: +17
         +9
Answer: +26

Example H
Subtract: –17
         +9
Solution: –17
         –9
Answer: –26

EXERCISE 29/Basic Time–9 Minutes

Estimated time to obtain a basic score of 100.

Divide these signed numbers: (8 each; 204 points)

1. $\frac{+35}{+7} =$
2. $\frac{-48}{-8} =$
3. $\frac{-54}{+9} =$

4. $\frac{-75}{+5} =$
5. $\frac{108}{-9} =$
6. $\frac{-84}{-7} =$

7. $\frac{112}{7} =$
8. $\frac{-96}{-8} =$
9. $\frac{-108}{6} =$

10. $\frac{144}{8} =$
11. $\frac{-56}{7} =$
12. $\frac{-6}{-1} =$

Subtract these signed numbers: (9 each)

13. +16
    +9

14. +15
    –8

15. –18
    +14

16. –38
    –17

17. +27
    –51

18. –37
     37

19. –38
     46

20. –48
    –54

21. 74
   52

22. 64
   76

23. –72
    –95

24. +68
    –54

TEST 29

Name _____ Total Score – _____
Date _____ Basic Score 100
Basic Time–9 Minutes   Improvement Score _____

Divide these signed numbers: (6 each)

1. $\dfrac{+56}{+8} =$ 2. $\dfrac{+84}{-6} =$ 3. $\dfrac{-98}{-7} =$ 4. $\dfrac{-72}{+9} =$

5. $\dfrac{68}{-4} =$ 6. $\dfrac{-96}{-12} =$ 7. $\dfrac{-112}{8} =$ 8. $\dfrac{205}{-5} =$

9. $\dfrac{-92}{-4} =$ 10. $\dfrac{-126}{7} =$ 11. $\dfrac{135}{-5} =$ 12. $\dfrac{-162}{-6} =$

13. $\dfrac{119}{-7} =$ 14. $\dfrac{108}{-9} =$ 15. $\dfrac{-32}{8} =$ 16. $\dfrac{-15}{1} =$

Subtract these signed numbers: (8 Each)

17. +17   18. −12   19.   28   20. −36   21. −35
    +28   − 4   −40   56   32

22.   87   23.   23   24. −76   25. −30
    −97   58   −38   17

26.   20   27. −52   28. −18   29. −145
    −19   −84   18   −78

LESSON 30 — Solving Equations

The equation is a powerful means for solving problems. Solving an equation unravels a mystery—the mystery of the value of x or some other symbol that is used to represent a number in an equation. In solving an equation, the terms are rearranged so that the term representing the unknown quantity appears on one side of the equals sign and the numerals appear on the other side.

A popular method of solving equations is based on transposition. To *transpose* means to transfer a term from one side of the equals sign to the other. This method is based on the following principle:

When an addend or subtrahend is transferred from one side of the equals sign to the other, its sign is changed from positive to negative or from negative to positive.

The answer obtained in solving an equation should be checked by substituting that value in the equation in place of the symbol for the unknown quantity. If the answer is correct, the left side of the equation will truly equal the right side.

Example A Find the value of x in the equation $x + 7 = 23$.

Solution	Check
$x + 7 = 23$	$16 + 7 = 23$
$x = 23 - 7$	$23 = 23$
$x = 16$	

Example B Solve the equation $n - 8 = 32$.

Solution	Check
$n - 8 = 32$	$40 - 8 = 32$
$n = 32 + 8$	$32 = 32$
$n = 40$	

EXERCISE 30/Basic Time–10 Minutes

Estimated time to obtain a basic score of 100.

Solve these equations. After checking each answer, place it in the appropriate space on the right. (20 each)

1. $x + 23 = 51$
2. $n + 18 = 35$
3. $43 + y = 87$
4. $58 + z = 82$
5. $n - 36 = 64$
6. $x - 39 = 25$
7. $y + 78 = 37$
8. $46 + z = 30$
9. $42 = n - 24$
10. $73 = x + 56$

	Name	Total Score _____
TEST 30	Date	Basic Score 100
	Basic Time–10 Minutes	Improvement Score _____

Solve these equations. Check each answer and place it in the appropriate space on the right. (20 each)

1. $n + 16 = 78$
2. $x - 33 = 43$

1. _____

2. _____

3. $y - 7 = 54$
4. $z + 72 = 42$

3. _____

4. _____

5. $18 + n = 72$
6. $24 = x - 37$

5. _____

6. _____

7. $y + 54 = 35$
8. $44 + z = 59$

7. _____

8. _____

9. $79 = x + 48$
10. $55 = n - 39$

9. _____

10. _____

LESSON 31
Multiplying and Collecting Terms in Equations

One method of showing multiplication is to closely join the symbols. That is, $5n$ means 5 *times* n, and $4(x + 1)$ means 4 *times* the quantity of $x + 1$.

When a multiplier or divisor is transferred from one side of an equals sign to the other side, its sign is changed from × to ÷ or from ÷ to ×.

Notice in Examples A and B that the symbol being transferred is placed to the right of the symbols already on that side of the equals sign.

Example A Solve this equation: $z ÷ 5 = 34$
Solution: $z ÷ 5 = 34$ **Check:** $170 ÷ 5 = 34$
$z = 34 × 5$ $34 = 34$
$z = 170$

Example B Solve this equation: $10n - 42 = 4n + 48$
Solution: **Check:**
$10n - 42 = 4n + 48$ $(10 × 15) - 42 = (4 × 15) + 48$
$10n - 4n = 48 + 42$ $150 - 42 = 60 + 48$
$6n = 90$ $108 = 108$
$n = 90 ÷ 6$
$n = 15$

EXERCISE 31/Basic Time—18 Minutes
Estimated time to obtain a basic score of 100.

Solve these equations. Check each answer and place it in the appropriate space on the right. (20 each)

1. $9n = 369$
2. $15x = 780$
3. $y ÷ 7 = 91$
4. $z ÷ 9 = 72$
5. $3x + 6 = 18$
6. $3x - 7 = 41$
7. $5y - 75 = 3y + 17$
8. $86 - 7z = z + 46$
9. $3x + x + 2x = 105 - x$
10. $3n - 2 = 5n - 3n + 32$

1. _____
2. _____
3. _____
4. _____
5. _____
6. _____
7. _____
8. _____
9. _____
10. _____

© Copyright South-Western Publishing Co. **65**

TEST 31

Name _____ Total Score _____
Date _____ Basic Score __100__
Basic Time–17 Minutes   Improvement Score _____

Solve these equations. Check each answer and place it in the appropriate space on the right. (20 each)

1. $22x = 748$

2. $41n = 1{,}722$

3. $y \div 65 = 8$

4. $8z + 16 = 760$

5. $7n - 27 = 421$

6. $50 + 4x = 414$

7. $8n + 94 = 3n + 124$

8. $10x - 23 = 6x + 77$

9. $3 + 5y + 6 + y = 129 - 4y$

10. $9z + 15 = 7z - 25 + 4z$

1. _____

2. _____

3. _____

4. _____

5. _____

6. _____

7. _____

8. _____

9. _____

10. _____

LESSON 32 Order of Operations

The order in which the arithmetic operations should be done when solving an equation are as follows: (1) Do all operations that are in parentheses or in other signs of aggregation, such as brackets or braces. (2) Do all multiplications and divisions as they occur from left to right. (3) Do all additions and subtractions as they occur from left to right.

In Example A the parentheses show that the 11 and 2 are grouped together, so the operation of addition is done first.

In Example B the product of the first three factors is to be divided by the product of 3 × 5, or divided by the 3 and the result divided by 5.

Example C illustrates all three of the rules for the order of operations.

1. Do operations within parentheses first and eliminate the parentheses.
2. Multiply and divide from left to right.
3. Add and subtract from left to right.

Example A $x = 23 - (11 + 2)$
$x = 23 - 13$
$x = 10$

Example B $n = 2 \times 7 \times 75 \div (3 \times 5)$
$n = 1{,}050 \div 15$ [or $1{,}050 \div 3 \div 5$]
$n = 70$

Example C $y = 26 - 8 \times 12 \div (6 - 2) + 9$
$y = 26 - 8 \times 12 \div 4 + 9$
$y = 26 - 96 \div 4 + 9$
$y = 26 - 24 + 9$
$y = 11$

EXERCISE 32/Basic Time – 20 Minutes
Estimated time to obtain a basic score of 100.

For each of the following, indicate which operation should be done first. Place each answer in the appropriate space. (10 each)

1. $(30 - 6) \times 3$
2. $30 + 6 \div 3$
3. $(30 + 6) \div 3$
4. $32 - 2 \times 4$
5. $25 + 7 \times 4$

Solve each of the following equations. Place each answer in the appropriate space. (10 each)

6. $n = 288 \div 36 \times 3$
7. $y = 18 \times 30 + 24 \times 12$
8. $x = 720 \div (45 \times 2)$
9. $a = 35 + 28 \times 21 - 14$
10. $z = 30 \times 16 - 16 + 56$
11. $a = 72 \times 40 - 48 \div 24$
12. $b = 48 \times 36 + 72 \div 24$
13. $c = 224 - (140 \div 35)$
14. $x = 190 - (117 \div 13)$
15. $y = (104 \times 130 - 26) \div 78$
16. $n = (450 + 30 \times 75) \div 150$
17. $x = (24 \times 140 - 12 \times 168) \div (140 - 44)$
18. $z = (81 \times 54 + 261 \times 108) \div (135 + 108)$
19. $y = (160 \div 40 + 28 \times 8) \div (3 \times 4)$
20. $n = (392 \div 8 + 35 \times 14) \div (7 \times 7)$

TEST 32

Name _____   Total Score _____

Date _____   Basic Score 100

Basic Time–22 Minutes   Improvement Score _____

Complete each of the following statements by supplying the missing word(s). Place each answer in the appropriate space on the right. (8 each)

1. Regarding the order of arithmetic operations, after all signs of aggregation are eliminated, the operations of __?__ and __?__ should be done.

    1. _____

2. The final step in the rules for the order of arithmetic operations is to do all __?__ and __?__ as they occur from left to right.

    2. _____

3. In $(20 - 10) \div 2$, the first arithmetic operation to perform is __?__.

    3. _____

4. To find the value of x in $x = 5 - (10 + 3)$, the first arithmetic operation to perform is __?__.

    4. _____

5. To solve the equation $n = 40 - 9 + 3$, the arithmetic operation you should do first is __?__.

    5. _____

6. In $z = 20 - 8 \div 4$, the arithmetic operation to do first is __?__.

    6. _____

7. In $32 + 2 \times 4$, the arithmetic operation to do first is __?__.

    7. _____

8. To solve the problem $n = 20 + 12 \div 4$, the arithmetic operation that should be done first is __?__.

    8. _____

9. In the problem $y = 14 - (5 + 3)$, the arithmetic operation that should be done first is __?__.

    9. _____

10. In the problem $y = 30 + 5 \times 2$, the arithmetic operation that should be done first is __?__.

    10. _____

Solve each of the following equations. (8 each)

11. $a = 54 + 32 \times 9 - 12$

12. $n = 18 \times 30 + 24 \div 12$

    11. _____

    12. _____

13. $y = 17 \times 20 + 35 \times 14$

14. $b = 21 \times 40 - 72 \div 24$

    13. _____

    14. _____

15. $x = (22 \times 14 - 12 \times 16) \div 29$

16. $y = (81 \times 53) - (46 \times 39)$

    15. _____

    16. _____

17. $z = (192 \times 64) \div (48 \times 16)$

18. $c = 146 - (280 \div 14)$

    17. _____

    18. _____

19. $x = 324 \div (36 \times 3)$

20. $a = 36 + 12 \times 24 - 64$

    19. _____

    20. _____

21. $n = (140 \times 32 - 16) \div 279$

    21. _____

22. $x = (160 \div 40 + 32 \times 16) \div (27 + 16)$

    22. _____

23. $y = (280 - 30 + 20 \times 55) \div (46 - 23 + 27)$

    23. _____

24. $x = (46 \times 82) + (93 \times 58)$

25. $n = (126 \times 84) \div (21 \times 14)$

    24. _____

    25. _____

68 © Copyright South-Western Publishing Co.

LESSON 33 — Parentheses in Equations

Parentheses are necessary in some equations to show which symbols are to be treated as one quantity. For example, $24 \div 2 + 4$ gives $12 + 4 = 16$; but, $24 \div (2 + 4)$ gives $24 \div 6 = 4$. The parentheses indicate that the 2 and 4 are associated (grouped together), so the addition should be done before the division. Parentheses and other signs of aggregation in an equation should be eliminated as soon as possible.

When parentheses that enclose addends and subtrahends are preceded by a factor, each symbol within the parentheses is multiplied by the factor. *Remember, a negative times a negative gives a positive.*

The example shows the solution of the equation $5(n + 3) - 2(n - 4) = 173$.

Example
$$5(n + 3) - 2(n - 4) = 173$$
$$5n + 15 - 2n + 8 = 173$$
$$5n - 2n = 173 - 15 - 8$$
$$3n = 150$$
$$n = 150 \div 3$$
$$n = 50$$

Check
$$5(50 + 3) - 2(50 - 4) = 173$$
$$5(53) - 2(46) = 173$$
$$265 - 92 = 173$$
$$173 = 173$$

EXERCISE 33/Basic Time – 24 Minutes
Estimated time to obtain a basic score of 100.

Solve these equations. Check each answer and place it in the appropriate space on the right. (10 each for Problems 1-6; 14 each for Problems 7-16)

1. $5(n + 6) = 65$
2. $3(x - 5) = x + 17$
3. $4(30 - 4y) = 4y + 500$
4. $16z = 4(20 - z)$
5. $16 - 2(a - 4) = a - 24$
6. $7b \div (6 - 4) = 63$
7. $14 + 3(x + 6) = 4 + 2(3x + 5)$
8. $10 + 5(y + 4) = 4(y + 3) + 8$
9. $148 - 4(n - 3) = 16$
10. $2(a - 4) + 3(a - 2) = 151$
11. $3(b + 2) - 4(b - 3) = 50$
12. $3(c - 3) - 2(c - 2) = 120$
13. $4(5n - 2) = 152 - 6(n + 5)$
14. $3(5x - 1) = 2(3x + 6) + 417$
15. $2(3y - 4) = 340 - 3(y - 4)$
16. $5(z + 3) - (4z - 8) = 130$

1. _____
2. _____
3. _____
4. _____
5. _____
6. _____
7. _____
8. _____
9. _____
10. _____
11. _____
12. _____
13. _____
14. _____
15. _____
16. _____

TEST 33

Name _____ Total Score _____
Date _____ Basic Score __100__
Basic Time–22 Minutes Improvement Score _____

Solve these equations. Check each answer and place it in the appropriate space on the right. (20 each)

1. $7(n + 2) = 105$

2. $6(x - 4) = 5x + 30$

3. $60 - 5(y - 4) = y - 16$

4. $6 + 4(z + 2) = 2(z + 7) + 40$

5. $3(a + 4) - 6(a - 3) = 87$

6. $5(3b - 2) = 4(2b + 3) + 34$

7. $6(2c + 5) = 4(c - 20) - 2$

8. $5(5x + 20) = 3(x + 48)$

9. $5(y - 3) + 8 = 59 - (30 - y)$

10. $14z - 4(z + 2) = 2(z - 2) + 3(z - 8)$

1. _____
2. _____
3. _____
4. _____
5. _____
6. _____
7. _____
8. _____
9. _____
10. _____

LESSON 34 Algebraic Expressions

To change from words to algebraic terms, follow these two rules: (1) Select a symbol such as x to represent one of the unknown quantities. (2) Express each of the other unknown quantities in relation to that symbol.

Example A Candy costs n dollars a pound. Express each of the following:

Solutions:
The cost of 5 pounds: $5n$
The cost of x pounds: xn
How much $3 will buy: $3 \div n$
The cost of 4 pounds if price reduced $1 a pound: $4(n-1)$

Example B Mary is now x years old. Express her age for each of the following:

Solutions:
Her age 3 years ago: $x-3$
Her age 5 years from now: $x+5$
Her age y years ago: $x-y$
Her age n years from now: $x+n$
Twice her present age: $2x$
Half her present age: $x \div 2$

EXERCISE 34/Basic Time—25 Minutes

Estimated time to obtain a basic score of 100.

Use algebraic symbols to write each of the following. Place each answer in the appropriate space. (8 each)

1. a plus b
2. 6 more than x
3. r times b
4. p divided by r
5. One third of n
6. x decreased by 6
7. 4 times the sum of a and b
8. y increased by one half of itself

Let x represent the length in feet of a certain board. Write an algebraic expression that represents the length of a board that is: (3 each)

9. 4 feet longer
10. 3 feet shorter
11. Twice as long
12. Half as long

If y represents Ruth's present age in years, what algebraic expression represents her age: (3 each)

13. 7 years from now?
14. 5 years ago?
15. b years ago?
16. c years from now?

Solve these problems. Place each answer in the appropriate space on the right. (16 each)

17. The difference between two numbers is 23. If the smaller number is represented by s, what algebraic expression represents the other number?

18. The sum of two numbers is represented by x. If one of the numbers is 34, what algebraic expression represents the other number?

19. Show the cost of 3 pounds of candy at z dollars a pound.

20. Write an algebraic expression to show four times the sum of y and five.

21. The letter n represents an odd number. Write an algebraic expression to represent the next consecutive odd number.

22. Write an algebraic expression that represents the difference between 48 and twice the sum of x and four.

23. Candy sells for $8 a pound. Show an algebraic expression to represent the change returned if a $20 bill is given to pay for n pounds.

Name _____ Total Score _____
Date _____ Basic Score 100
Basic Time–22 Minutes Improvement Score _____

Use algebraic symbols to write each of the following. Place each answer in the appropriate space on the right. (10 each problem)

1. 7 added to x
2. 5 more than n
3. 8 less than y
4. 4 times z
5. n divided by 9
6. p times r
7. One half of x
8. 1.5 times y
9. 6 times the sum of z and 4
10. n decreased by one third of itself
11. A computer (Brand XYZ) sells for x dollars and another (Brand ABC) sells for $800 more. Express the selling price of the Brand ABC computer.
12. Write an algebraic expression for the number of months in y years.
13. There were 200 gallons of milk in a tank. How many gallons were in the tank after z gallons of milk were added to the tank?
14. The price of one dozen ballpoint pens is n dollars. Express the price of one pen.
15. One pencil costs x cents. Express the cost of one dozen pencils.
16. Ida Williams earns d dollars a month and her brother, Leroy, earns $200 a month more. Express Leroy's (a) monthly earnings and (b) annual earnings.
17. An employee earns $500 a week and spends an average of x dollars a day. How much can this employee save in 8 weeks?
18. Richard Doty invested $50,000 in two investments. If n represents the number of dollars placed in the first investment, what expression represents the other?
19. (a) The letter x represents an odd number. Write an algebraic expression to represent the next consecutive odd number. (b) The letter n represents an even number. Write an algebraic expression to represent the next consecutive even number.
20. The selling price of one doughnut is x cents and of one breakfast roll is y cents. Express the total selling price of 6 doughnuts and 12 rolls.

1. _____
2. _____
3. _____
4. _____
5. _____
6. _____
7. _____
8. _____
9. _____
10. _____
11. _____
12. _____
13. _____
14. _____
15. _____
16a. _____
16b. _____
17. _____
18. _____
19a. _____
19b. _____
20. _____

LESSON 35

Application Problems

There is no magic formula that can be used to guarantee finding the correct answer to every word problem. Following the rules below, however, can help you to solve most word problems.
1. Read the problem all the way through.
2. Determine what the problem asks.
3. Let a symbol such as x represent one of the unknown quantities.
4. Express the other unknown quantities in terms of this symbol.
5. Write an equation expressing the relationship between the given and the unknown quantities.
6. Solve the equation to find the value of the unknown quantities.
7. Check the solution.

EXERCISE 35/Basic Time–30 Minutes
Estimated time to obtain a basic score of 100.

Use an equation in good form to solve each of the following problems. Check each answer and write it in the appropriate space on the right. (10 each answer)

1. The sum of two numbers is 87. The smaller number is 15 less than the larger number. What is the larger number?

2. The sum of three consecutive even numbers is 822. Find the numbers.

3. The sum of three numbers is 165. The second number is twice the first, and the third is 15 less than the first. Find the numbers.

4. The sum of three consecutive odd numbers is 129. Find the numbers.

5. If twice a certain number is increased by 40, the result is the same as when four times the number is increased by 6. Find the number.

6. The second of two numbers is four less than three times the first. The sum of the numbers is 168. Find the numbers.

7. The partnership agreement of Fletcher, Graves, and Harrington provides that when earnings and losses are distributed, Harrington is to get twice as much as Fletcher and that Graves is to get twice as much as Harrington. Earnings for the past year totaled $199,500. How much should go to (a) Fletcher, (b) Graves, and (c) Harrington?

8. Alice Pence paid $18,000 in state and federal income taxes for the past year. The federal tax was four times as large as the state tax. How much did she pay in (a) state income tax and (b) federal income tax?

9. Cole and Davis invested $78,000 in a business. Davis invested $5,000 more than Cole. How much was invested by (a) Cole and (b) Davis?

10. When 79 is subtracted from three times a certain number, the result is the same as when 18 is subtracted from twice the number. Find the number.

11. Partners Brown and Campbell cleared $19,000 in a certain business transaction. By agreement Brown received three times as much as Campbell. How much should be distributed by the accountant to (a) Campbell and (b) Brown?

12. Mary drove 843 miles in two days. On the first day she drove 75 miles farther than she did the second day. How far did she drive the first day?

13. Ruemmler and Simmons invested $67,500 in a business. Ruemmler invested $3,500 more than Simmons. How much did Simmons invest in the business?

14. Thirty bills of currency consisting of $5 bills and $10 bills totaled $235. Find the number of (a) $5 bills and (b) $10 bills.

TEST 35

Name _____ Total Score _____
Date _____ Basic Score __100__
Basic Time–30 Minutes Improvement Score _____

Use an equation in good form to solve each of the following problems. Check each answer and write it in the appropriate space on the right. (10 each problem)

1. A number increased by 23 is 85. What is the number?
2. A number decreased by 35 is 97. What is the number?
3. Forty more than six times a number is 214. Find the number.
4. Nine less than five times a number is 276. Find the number.
5. When 19 is subtracted from a number, the result is 89. What is the number?
6. The sum of three numbers is 156. The second number is 4 more than the first, and the third is 14 more than the first. Find the numbers.
7. Ruth has $25 more than Susan. Together they have $247. How many dollars does Susan have?
8. Six times a certain number equals the number increased by 65. What is the number?
9. The sum of three consecutive odd numbers is 237. Find the numbers.
10. Lew is three years older than David. The sum of their ages is 49. How old are (a) David and (b) Lew?
11. If four times a certain number is increased by 47, the result is the same as when seven times the number is decreased by 34. Find the number.
12. The second of two numbers is eight less than four times the first. The sum of the numbers is 267. Find the numbers.
13. The partnership agreement of Keplar, Lowery, and Mundell provides that when earnings and losses are distributed, Mundell is to get three times as much as Keplar and that Lowery is to get twice as much as Mundell. Earnings for the past year totaled $240,000. How much should go to (a) Keplar, (b) Lowery, and (c) Mundell?
14. A house and lot together cost $267,500. The house cost $137,500 more than the lot. Find the cost of (a) the lot and (b) the house.
15. Allen and Baker invested $68,000 in a business. Baker invested $7,000 more than Allen. How much was invested by (a) Allen and (b) Baker?
16. When 93 is added to five times a certain number, the result is the same as when 29 is added to seven times the number. Find the number.
17. When they entered into partnership, Edwards and Frailey agreed that Frailey would receive twice as much as Edwards when gains and losses were distributed. How much of a loss of $123,540 should go to (a) Edwards and (b) Frailey?
18. A coat and dress together cost $285. The coat cost $97 more than the dress. How much did the dress cost?
19. Neal and Odum invested $67,500 in a business. Neal invested $5,700 more than Odum. How much did Odum invest in the business?
20. Seventy-one bills of currency consisting of $5 bills and $10 bills totaled $530. Find the number of (a) $5 bills and (b) $10 bills.

1. _____
2. _____
3. _____
4. _____
5. _____
6. _____
7. _____
8. _____
9. _____
10a. _____
10b. _____
11. _____
12. _____
13a. _____
13b. _____
13c. _____
14a. _____
14b. _____
15a. _____
15b. _____
16. _____
17a. _____
17b. _____
18. _____
19. _____
20a. _____
20b. _____

© Copyright South-Western Publishing Co.

2 Cumulative Review

Name _____

Complete each statement by supplying the missing word(s).

1. Numbers that use + or – to show they are positive or negative are called __?__ numbers.

   1._____

2. Numbers without a + or – sign are considered to be (positive/negative) __?__.

   2._____

3. To be considered negative, the – sign must be to the (left/right) __?__ of the number?

   3._____

4. In the multiplication of signed numbers, a positive number times a negative number gives a (positive/negative) __?__ product.

   4._____

5. In the division of signed numbers, when the divisor and dividend have unlike signs, the quotient is (positive/negative) __?__.

   5._____

6. When an addend is transferred from one side of the equals sign to the other, the operation sign (is/is not) __?__ changed.

   6._____

7. When a multiplier is transferred from one side of the equals sign to the other, the operation sign (is/is not) __?__ changed?

   7._____

8. As a matter of form, terms being transferred from one side of the equals sign to the other should be written to the (left/right) __?__ of those terms already on the side to which the transfer is made.

   8._____

9. In the problem $y = 42 + 18 \div 3$, the arithmetic operation that should be done first is __?__.

   9._____

10. In the problem $y = 20 + 5 \times 2$, the arithmetic operation that should be done first is __?__.

    10._____

Add these signed numbers.

11. +7
     +6

    11._____

12. –9
     –7

    12._____

13. +7
     –5
     +6

    13._____

14. –86
      99
     –34

    14._____

15. –43
      57
     – 2
      34
     –26

    15._____

© Copyright South-Western Publishing Co.

2 Cumulative Review

Multiply these signed numbers.

16. 9
     × 6

17. 9
     × −6

18. −9
     × 6

19. −9
     × −6

20. −807
     ×   5

21. (−7)(6)(−9) =

22. (−4)(3)(−4)(2)(−8) =

Subtract these signed numbers.

23. −58
     − 32

24. −140
     − −95

25. 628
     − 409

26. 410
     − −268

Divide these signed numbers.

27. $\dfrac{52}{-4}$

28. $\dfrac{-72}{-8}$

29. $\dfrac{-68}{4}$

30. $\dfrac{351}{-9}$

16. _____
17. _____
18. _____
19. _____
20. _____
21. _____
22. _____
23. _____
24. _____
25. _____
26. _____
27. _____
28. _____
29. _____
30. _____

2 Cumulative Review

Name _____

Solve these equations. Check your answers.

31. $n + 15 = 51$

32. $x - 17 = 72$

33. $4y + 3 = 75$

34. $6x + 16 = 5x + 3x - 62$

35. $n = 810 \div (45 \times 2)$

36. $x = 36 \times 10 - 72 \div 12$

37. $a = (45 + 3 \times 25) \div 5$

38. $b = (320 \div 20 + 41 \times 4) \div 5 \times 3$

39. $4(x - 3) + 2(x - 1) = 106$

40. $5(z + 4) - 3(z - 3) = 39$

41. $3(7n - 3n - 6) = 5(n - 2) + 6(n + 1) - 2$

Use algebraic symbols to write each of the following.

42. One half of n

43. x less than 7

44. y decreased by $\frac{1}{3}$

45. z more than 9

46. 3 times the quantity of n increased by 6

47. A computer sells for y dollars. A different model sells for $800 more. Express the selling price of the second computer.

48. Let x represent Joan's present annual salary. Write an algebraic expression to represent each of the following conditions.

   a. At this salary, how much money will she earn in two years.
   b. If her salary is increased by $6,000 next year, how much will her new salary be?
   c. What represents her monthly salary?
   d. How much is her salary for one-half year?
   e. If her salary is decreased by $400 a month how much will her new salary be?

Use representative letters to write an equation for each of these.

49. The rate equals the percentage divided by the base.

50. The cost of a fixed asset, such as a machine, less its accumulated depreciation gives the book value of the fixed asset.

51. The sales tax rate times the selling price of an item gives the amount of the sales tax on the item.

52. A worker's gross pay is found by multiplying the hours worked by the hourly rate of pay.

31. _____
32. _____
33. _____
34. _____
35. _____
36. _____
37. _____
38. _____
39. _____
40. _____
41. _____
42. _____
43. _____
44. _____
45. _____
46. _____
47. _____
48a. _____
48b. _____
48c. _____
48d. _____
48e. _____
49. _____
50. _____
51. _____
52. _____

2 Cumulative Review

Solve these word problems.

53. If *A* represents assets and *E* equities in the equation $A = \$45,000 + E$, the assets are how much larger than the equities.

53. _____

54. The sum of two numbers is 326. One of the numbers is 48 larger than the other number. Find the numbers.

54. _____

55. The sum of three numbers is 321. The second number is twice the first number, and the third is 6 more than the second number. Find the numbers.

55. _____

56. The sum of two numbers is 34. Four times the larger number is 4 more than 7 times the smaller number. Find the numbers.

56. _____

57. Alicia traveled 859 miles in two days. On the second day, she traveled 63 miles farther than on the first day. How many miles did she travel the first day?

57. _____

58. Garcia and Huntington invested $134,500 in a business. Garcia invested $14,500 more than Huntington. How much did Garcia invest?

58. _____

59. The partnership contract of Chiba, Diaz, and Enami provides that Diaz is to receive twice as much as Chiba and Enami three times as much as Diaz when gains and losses are distributed. How much of $126,000 in earning should go to (a) Chiba, (b) Diaz, and (c) Enami?

59a. _____
59b. _____
59c. _____

60. After the taxes, legal fees, and other claims had been paid, $132,480 remaining of an estate was divided among the wife, two children, and two grandchildren of the deceased. The wife received three times as much as each child; and each child, twice as much as each grandchild.

a. How much was received by each grandchild?

b. By each child?

c. By the wife?

60a. _____
60b. _____
60c. _____

LESSON 36 Common Denominator

A part of a whole number is called a *fraction*. A fraction expresses a relationship between two numbers. The top number of a fraction is the *numerator*. The bottom number is the *denominator*.

In addition and subtraction, only like denominators should be added or subtracted. In the addition of fractions, if the denominators are different, the fractions must be changed to equivalent fractions with a *common denominator;* that is, all denominators must be the same.

To change a fraction to its equivalent in a higher denomination, multiply the denominator by the number that will give the desired denominator; then multiply the numerator by the same number. See Example A.

To change a fraction to its equivalent in a lower denomination, divide both the numerator and denominator by the same number. See Example B.

Example A: Change to Higher Terms
$$\frac{4}{5} = \frac{4 \times 4}{5 \times 4} = \frac{16}{20}$$

Example B: Change to Lower Terms
$$\frac{3}{12} = \frac{3 \div 3}{12 \div 3} = \frac{1}{4}$$

EXERCISE 36/Basic Time – 2 Minutes
Estimated time to obtain a basic score of 100.

	a	b	c	d	
Change to 12ths:					
1.	$\frac{1}{2} = \frac{\ }{12}$	$\frac{2}{3} = \frac{\ }{12}$	$\frac{3}{4} = \frac{\ }{12}$	$\frac{5}{6} = \frac{\ }{12}$	(6)
Change to 32ds:					
2.	$\frac{3}{4} = \frac{\ }{32}$	$\frac{5}{8} = \frac{\ }{32}$	$\frac{7}{8} = \frac{\ }{32}$	$\frac{9}{16} = \frac{\ }{32}$	(6)
Change to 64ths:					
3.	$\frac{3}{8} = \frac{\ }{64}$	$\frac{3}{4} = \frac{\ }{64}$	$\frac{5}{16} = \frac{\ }{64}$	$\frac{7}{32} = \frac{\ }{64}$	(12)
Change to 4ths:					
4.	$\frac{12}{16} = \frac{\ }{4}$	$\frac{18}{24} = \frac{\ }{4}$	$\frac{24}{32} = \frac{\ }{4}$	$\frac{28}{56} = \frac{\ }{4}$	(8)
Change to 8ths:					
5.	$\frac{15}{24} = \frac{\ }{8}$	$\frac{44}{88} = \frac{\ }{8}$	$\frac{20}{32} = \frac{\ }{8}$	$\frac{42}{56} = \frac{\ }{8}$	(8)
Change to 16ths:					
6.	$\frac{15}{48} = \frac{\ }{16}$	$\frac{25}{80} = \frac{\ }{16}$	$\frac{28}{64} = \frac{\ }{16}$	$\frac{18}{32} = \frac{\ }{16}$	(10)

© Copyright South-Western Publishing Co.

TEST 36

Name ___ Total Score ___
Date ___ Basic Score 100
Basic Time–2 Minutes Improvement Score ___

Change to higher or lower terms as indicated.

        a   b   c   d

1. $\frac{1}{2} = \frac{}{8}$ $\frac{1}{3} = \frac{}{12}$ $\frac{1}{4} = \frac{}{16}$ $\frac{1}{6} = \frac{}{72}$ (4)

2. $\frac{2}{3} = \frac{}{48}$ $\frac{3}{4} = \frac{}{32}$ $\frac{5}{6} = \frac{}{24}$ $\frac{7}{8} = \frac{}{64}$ (6)

3. $\frac{11}{15} = \frac{}{75}$ $\frac{13}{16} = \frac{}{64}$ $\frac{19}{24} = \frac{}{72}$ $\frac{17}{32} = \frac{}{96}$ (8)

4. $\frac{3}{16} = \frac{}{48}$ $\frac{5}{9} = \frac{}{63}$ $\frac{7}{12} = \frac{}{48}$ $\frac{9}{16} = \frac{}{64}$ (7)

5. $\frac{4}{16} = \frac{}{4}$ $\frac{6}{8} = \frac{}{4}$ $\frac{12}{16} = \frac{}{4}$ $\frac{12}{15} = \frac{}{5}$ (4)

6. $\frac{15}{25} = \frac{}{5}$ $\frac{20}{24} = \frac{}{6}$ $\frac{24}{36} = \frac{}{3}$ $\frac{18}{30} = \frac{}{5}$ (6)

7. $\frac{20}{32} = \frac{}{8}$ $\frac{14}{21} = \frac{}{3}$ $\frac{15}{36} = \frac{}{12}$ $\frac{21}{24} = \frac{}{8}$ (4)

8. $\frac{30}{36} = \frac{}{6}$ $\frac{15}{36} = \frac{}{12}$ $\frac{21}{56} = \frac{}{8}$ $\frac{20}{45} = \frac{}{9}$ (4)

9. $\frac{3}{8} = \frac{}{72}$ $\frac{5}{25} = \frac{}{5}$ $\frac{7}{16} = \frac{}{80}$ $\frac{13}{52} = \frac{}{4}$ (7)

LESSON 37
Lowest Terms of Common Fractions

Numbers such as $\frac{1}{2}$, $\frac{3}{4}$, and $\frac{10}{12}$ in which the numerators are smaller than the denominators, are called *common fractions*. It is necessary to know how to reduce common fractions to lowest terms when they are not already so expressed. To change a common fraction to its lowest terms, divide the numerator and the denominator by all the divisors that are common to both. See Example A. When a common divisor cannot be seen readily, find the *greatest common divisor* (*gcd*). See Example B.

Example A
$$\frac{18}{24} = \frac{18 \div 2}{24 \div 2} = \frac{9}{12}; \frac{9}{12} = \frac{9 \div 3}{12 \div 3} = \frac{3}{4}$$

Example B Change $\frac{39}{65}$ to its lowest terms.

Solution $\frac{39 \div 13}{65 \div 13} = \frac{3}{5}$

```
        1
    39)65
       39        1
       26)39
          26       2
          13)26
             26
              0    gcd
```

Divide the larger number by the smaller. Then divide the remainder into the preceding divisor. Continue this procedure until there is no remainder. The *last divisor* is the greatest common divisor (gcd).

EXERCISE 37/Basic Time–4 Minutes

Estimated time to obtain a basic score of 100.

Find the gcd and change each fraction to lowest terms: (198 points)

	a	b	c
1.	$\frac{12}{16}$ = _____ gcd _____	$\frac{24}{64}$ = _____ gcd _____	$\frac{18}{32}$ = _____ gcd _____ (8) (2)
2.	$\frac{9}{24}$ = _____ gcd _____	$\frac{17}{30}$ = _____ gcd _____	$\frac{35}{42}$ = _____ gcd _____ (8) (2)
3.	$\frac{36}{84}$ = _____ gcd _____	$\frac{15}{60}$ = _____ gcd _____	$\frac{29}{96}$ = _____ gcd _____ (12) (2)
4.	$\frac{57}{95}$ = _____ gcd _____	$\frac{91}{156}$ = _____ gcd _____	$\frac{51}{85}$ = _____ gcd _____ (14) (2)
5.	$\frac{46}{115}$ = _____ gcd _____	$\frac{58}{87}$ = _____ gcd _____	$\frac{63}{155}$ = _____ gcd _____ (14) (2)

TEST 37

Name _____ Total Score _____

Date _____ Basic Score __100__

Basic Time–3 Minutes Improvement Score _____

Change to lowest terms: (198 points)

         a   b   c

1. $\dfrac{15}{36} =$ $\dfrac{42}{66} =$ $\dfrac{18}{54} =$ (10)

2. $\dfrac{46}{75} =$ $\dfrac{24}{40} =$ $\dfrac{63}{81} =$ (10)

3. $\dfrac{28}{98} =$ $\dfrac{90}{135} =$ $\dfrac{56}{104} =$ (14)

4. $\dfrac{32}{144} =$ $\dfrac{48}{116} =$ $\dfrac{27}{108} =$ (16)

5. $\dfrac{39}{91} =$ $\dfrac{48}{84} =$ $\dfrac{48}{92} =$ (16)

LESSON 38
Improper Fractions and Mixed Numbers

Numbers such as $2\frac{1}{2}$, $5\frac{3}{4}$, and $6\frac{2}{5}$, which are composed of a whole number and a fraction, are *mixed numbers*. Fractions such as $\frac{4}{4}$, $\frac{7}{5}$, and $\frac{8}{3}$, in which the numerator is as large as or larger than the denominator, are *improper fractions*.

Every mixed number can be expressed as an improper fraction. To change a mixed number to an improper fraction, multiply the whole number by the denominator and to this product add the numerator. This final number will be the numerator of the improper fraction. The denominator will not change. See Example A.

To change an improper fraction to a mixed number, divide the numerator by the denominator. The quotient will be the whole number. The remainder (if there is one) will be placed over the denominator of the original improper fraction to form the fractional part of the mixed number. See Example B.

Example A Change to Improper Fraction

$$4\frac{3}{5} = \frac{(4 \times 5) + 3}{5} = \frac{23}{5}$$

Example B Change to Mixed Number

$$\frac{23}{5} = 5\overline{)23} = 4\frac{3}{5}$$

EXERCISE 38/Basic Time – 3 Minutes

Estimated time to obtain a basic score of 100.

Change to improper fractions: (198 points)

	a	b	c	
1.	$2\frac{3}{4} =$	$5\frac{7}{8} =$	$6\frac{5}{8} =$	(3)
2.	$9\frac{11}{12} =$	$12\frac{7}{9} =$	$8\frac{9}{16} =$	(5)
3.	$15\frac{17}{20} =$	$21\frac{15}{16} =$	$25\frac{7}{15} =$	(12)
4.	$3\frac{19}{64} =$	$5\frac{17}{48} =$	$23\frac{11}{27} =$	(12)

Change to mixed numbers:

	a	b	c	
5.	$\frac{17}{5} =$	$\frac{35}{8} =$	$\frac{47}{12} =$	(4)
6.	$\frac{53}{6} =$	$\frac{55}{12} =$	$\frac{49}{9} =$	(4)
7.	$\frac{121}{64} =$	$\frac{134}{32} =$	$\frac{156}{23} =$	(12)
8.	$\frac{541}{16} =$	$\frac{642}{9} =$	$\frac{463}{24} =$	(14)

© Copyright South-Western Publishing Co.

TEST 38

Name
Date
Basic Time–3 Minutes

Total Score _____
Basic Score __100__
Improvement Score _____

Change to improper fractions: (198 points)

	a	b	c	
1.	$7\frac{1}{4}=$	$6\frac{3}{8}=$	$4\frac{8}{9}=$	(3)
2.	$8\frac{5}{12}=$	$9\frac{2}{11}=$	$7\frac{3}{16}=$	(4)
3.	$13\frac{11}{16}=$	$12\frac{5}{24}=$	$14\frac{13}{15}=$	(14)
4.	$5\frac{19}{64}=$	$6\frac{17}{48}=$	$13\frac{5}{17}=$	(14)

Change to mixed numbers:

5.	$\frac{18}{5}=$	$\frac{27}{8}=$	$\frac{35}{12}=$	(4)
6.	$\frac{62}{7}=$	$\frac{53}{16}=$	$\frac{72}{15}=$	(5)
7.	$\frac{73}{32}=$	$\frac{137}{64}=$	$\frac{169}{48}=$	(8)
8.	$\frac{657}{16}=$	$\frac{329}{24}=$	$\frac{427}{64}=$	(14)

LESSON 39: Lowest Common Denominator

A number that is divisible, without a remainder, by a set of numbers is a *common multiple* of that set. A number larger than 1 that is divisible only by 1 and by itself is a *prime number*. For example, 2, 3, 5, 7, 11, and 13 are the six smallest prime numbers.

A set of numbers will have many common multiples. The smallest of these is the *lowest common multiple (lcm)*. To find the *lcm* of a set of numbers, arrange the numbers in a horizontal row and divide by any prime number common to two or more of the numbers of the set. Carry down each quotient together with any number that is not divisible by the chosen prime number. Continue dividing in this manner until no two numbers can be divided by a prime number.

Then multiply together all the prime-number divisors and the final quotients.

The lowest common multiple of the denominators in a set of fractions is the *lowest common denominator (lcd)*.

Example Find the lowest common multiple (lcm) of 8, 16, 20, and 25.

Solution:
```
2)8   16   20   25
2)4    8   10   25
2)2    4    5   25
5)1    2    5   25
  1    2    1    5  ← final quotients
```
$2 \times 2 \times 2 \times 5 \times 1 \times 2 \times 1 \times 5 = 400$ **lcm**

400 is the smallest number into which 8, 16, 20, and 25 will all divide.

EXERCISE 39/Basic Time—8 Minutes
Estimated time to obtain a basic score of 100.

Find the lowest common multiple of each set of numbers. Place each answer in the appropriate space on the right.

1. 8, 12, 15, 20
2. 16, 20, 24, 40

1. _____ (20)

2. _____ (20)

3. 25, 35, 42, 45
4. 12, 18, 25, 35

3. _____ (20)

4. _____ (20)

5. 14, 32, 35, 50
6. 32, 42, 48, 52

5. _____ (24)

6. _____ (24)

Find the lowest commond denominator of each set of fractions. Place each answer in the appropriate space on the right.

7. $\frac{5}{6}, \frac{7}{8}, \frac{5}{9}, \frac{7}{12}$
8. $\frac{8}{15}, \frac{9}{16}, \frac{13}{20}, \frac{11}{32}$

7. _____ (16)

8. _____ (16)

9. $\frac{9}{16}, \frac{7}{12}, \frac{5}{18}, \frac{11}{32}$
10. $\frac{5}{12}, \frac{15}{32}, \frac{11}{48}, \frac{19}{64}$

9. _____ (20)

10. _____ (20)

TEST 39

Name _____ Total Score _____

Date _____ Basic Score 100

Basic Time–7 Minutes Improvement Score _____

Find the lowest common multiple of each set of numbers. Place each answer in the appropriate space on the right.

1. 6, 8, 9, 12, 15

2. 15, 16, 18, 20, 24

1. _____ (20)

2. _____ (20)

3. 7, 8, 9, 10, 16

4. 20, 25, 50, 35, 40

3. _____ (20)

4. _____ (20)

Find the lowest common denominator of each set of fractions. Place each answer in the appropriate space on the right.

5. $\frac{7}{8}, \frac{5}{12}, \frac{3}{16}, \frac{17}{24}$

6. $\frac{14}{15}, \frac{11}{20}, \frac{17}{25}, \frac{1}{30}$

5. _____ (16)

6. _____ (16)

7. $\frac{13}{15}, \frac{15}{32}, \frac{17}{48}, \frac{7}{64}$

8. $\frac{5}{9}, \frac{15}{16}, \frac{11}{24}, \frac{17}{30}$

7. _____ (24)

8. _____ (24)

9. $\frac{5}{18}, \frac{7}{24}, \frac{11}{32}, \frac{19}{40}$

10. $\frac{11}{16}, \frac{17}{20}, \frac{5}{28}, \frac{15}{32}$

9. _____ (20)

10. _____ (20)

LESSON 40: Estimation with Fractions

Estimation is especially useful when computing with fractions. Completing this exercise will improve your understanding of fractional size. This will help you decide whether or not your answers to problems containing fractions "make sense."

EXERCISE 40/Basic Time–6 Minutes
Estimated time to obtain a basic score of 100.

Read the instructions carefully. (10 each problem)

1. Finish these fractions so that they are close to but greater than $\frac{1}{2}$.

   $\frac{}{7}$ $\quad \frac{}{9}$ $\quad \frac{}{12}$ $\quad \frac{}{16}$ $\quad \frac{7}{}$ $\quad \frac{8}{}$ $\quad \frac{12}{}$

2. Finish these fractions so that they are close to but smaller than 1.

   $\frac{}{8}$ $\quad \frac{}{11}$ $\quad \frac{}{16}$ $\quad \frac{9}{}$ $\quad \frac{12}{}$ $\quad \frac{6}{}$ $\quad \frac{15}{}$

3. The sum of $\frac{11}{12} + \frac{5}{6}$ must be (a) less than 1, (b) exactly 1, or (c) more than 1.   3. _____

4. The sum of $\frac{2}{8} + \frac{1}{5}$ must be (a) less than $\frac{1}{2}$, (b) exactly $\frac{1}{2}$, or (c) more than $\frac{1}{2}$.   4. _____

5. The sum of $\frac{14}{29} + \frac{13}{27} + \frac{7}{16} + \frac{3}{7}$ must be (a) less than 2, (b) exactly 2, or (c) more than 2.   5. _____

6. The sum of $\frac{7}{8} + \frac{4}{5} + \frac{11}{15} + \frac{3}{4}$ must be (a) less than 2, (b) exactly 2, or (c) more than 2.   6. _____

Show an estimate. Then show whether your estimate should be increased or decreased by circling the appropriate symbol. Do not adjust your estimate or compute the exact answer.

7. $\frac{4}{5} + \frac{1}{8} + \frac{1}{2} + \frac{5}{6}$ 8. $\frac{2}{5} + \frac{1}{4} + \frac{1}{2} + \frac{7}{84}$

9. $2\frac{2}{3} + \frac{3}{5} + 4\frac{1}{2} + 1\frac{3}{8}$ 10. $3\frac{1}{6} + 1\frac{5}{8} + \frac{5}{8} + 5\frac{3}{4}$

7. ____ + –
8. ____ + –
9. ____ + –
10. ____ + –

Show an estimate. Use a + or – to show if your estimate should be increased or decreased.

11. $\frac{7}{8} + \frac{3}{4} + \frac{11}{12} + \frac{1}{2}$ 12. $\frac{2}{3} + \frac{1}{4} + \frac{3}{8} + \frac{5}{12}$

13. $6\frac{1}{3} + \frac{2}{5} + 2\frac{2}{3} + 1\frac{3}{5}$ 14. $3\frac{1}{2} + \frac{7}{8} + 1\frac{3}{4} + 4\frac{1}{4}$

15. $2\frac{13}{15} + \frac{15}{16} + 3\frac{1}{12} + 4\frac{2}{5}$ 16. $5\frac{1}{3} + 7\frac{1}{5} + \frac{5}{8} + 2\frac{1}{3}$

11. _____
12. _____
13. _____
14. _____
15. _____
16. _____

Do not compute the exact answer. Show an estimate. Use a + or – to show if your answer should be increased or decreased.

17. $\frac{1}{4} \times 42$ 18. $\frac{2}{5} \times 164$

19. $\frac{1}{3} \times 50$ 20. $\frac{3}{4} \times 185$

17. _____
18. _____
19. _____
20. _____

Test 40

Name _____ Total Score _____
Date _____ Basic Score __100__
Basic Time–5 Minutes Improvement Score _____

Read the instructions carefully. (10 each problem)

1. Finish these fractions so that they are close to but greater than $\frac{1}{2}$.

   $\frac{}{6}$ $\frac{}{8}$ $\frac{5}{12}$ $\frac{4}{11}$ $\frac{5}{}$ $\frac{4}{}$ $\frac{10}{}$

2. Finish these fractions so that they are close to but smaller than 1.

   $\frac{}{7}$ $\frac{}{10}$ $\frac{}{14}$ $\frac{8}{}$ $\frac{11}{}$ $\frac{9}{}$ $\frac{13}{}$

3. The sum of $\frac{5}{12} + \frac{3}{7}$ must be (a) less than 1, (b) exactly 1, or (c) more than 1.   3. _____

4. The sum of $\frac{3}{8} + \frac{2}{5}$ must be (a) less than $\frac{1}{2}$, (b) exactly $\frac{1}{2}$, or (c) more than $\frac{1}{2}$.   4. _____

5. The sum of $\frac{15}{16} + \frac{4}{9} + \frac{9}{12} + \frac{3}{8}$ must be (a) less than 2, (b) exactly 2, or (c) more than 2.   5. _____

6. The sum of $\frac{1}{4} + \frac{1}{3} + \frac{2}{5} + \frac{2}{6}$ must be (a) less than 2, (b) exactly 2, or (c) more than 2.   6. _____

Show an estimate. Then show whether your estimate should be increased or decreased by circling the appropriate symbol. Do not adjust your estimate or compute the exact answer.

7. $\frac{3}{5} + \frac{7}{8} + \frac{3}{4} + \frac{5}{6}$ 8. $\frac{4}{5} + \frac{9}{10} + \frac{2}{3} + \frac{7}{8}$

9. $5\frac{2}{3} + \frac{3}{5} + 4\frac{1}{3} + 1\frac{3}{5}$ 10. $3\frac{1}{2} + 1\frac{3}{7} + \frac{5}{8} + 2\frac{3}{4}$

7. _____ + –
8. _____ + –
9. _____ + –
10. _____ + –

Show an estimate. Use a + or – to show if your estimate should be increased or decreased.

11. $\frac{7}{8} + \frac{3}{4} + \frac{7}{15} + \frac{1}{4}$ 12. $\frac{2}{3} + \frac{3}{4} + \frac{7}{8} + \frac{5}{12}$

13. $3\frac{1}{3} + \frac{2}{7} + 2\frac{2}{3} + 1\frac{3}{7}$ 14. $6\frac{1}{2} + \frac{7}{8} + 1\frac{3}{4} + 4\frac{1}{8}$

15. $1\frac{8}{9} + \frac{15}{16} + 3\frac{1}{6} + 2\frac{2}{5}$ 16. $5\frac{2}{3} + 4\frac{1}{5} + \frac{3}{10} + 1\frac{2}{3}$

11. _____
12. _____
13. _____
14. _____
15. _____
16. _____

Do not compute the exact answer. Show an estimate. Use a + or – to show if your estimate should be increased or decreased.

17. $\frac{1}{6} \times 32$ 18. $\frac{2}{3} \times 170$

19. $\frac{1}{4} \times 63$ 20. $\frac{3}{5} \times 127$

17. _____
18. _____
19. _____
20. _____

LESSON 41
Addition of Common Fractions

To add common fractions, all denominators must be the same. Therefore, change them to equivalent fractions with their lowest common denominator. Add the new numerators, and place this sum over the lowest common denominator. If the result is an improper fraction, change it to a mixed number in lowest terms.

When the lowest common denominator cannot be found through inspection, you can find it by following the procedure described in Exercise 39. In the example below, $\frac{2}{3}$ is added to $\frac{3}{4}$ by changing the two fractions to equivalent fractions with the lowest common denominator of 12.

Add: $\frac{2}{3} + \frac{3}{4} = \frac{8}{12} + \frac{9}{12} = \frac{17}{12} = 1\frac{5}{12}$

EXERCISE 41/Basic Time—8 Minutes
Estimated time to obtain a basic score of 100.

Change to equivalent fractions with common denominators and add:

1. $\frac{1}{4} + \frac{2}{3} =$
2. $\frac{3}{4} + \frac{5}{6} =$
3. $\frac{4}{5} + \frac{3}{4} =$
4. $\frac{7}{8} + \frac{5}{6} =$
5. $\frac{5}{6} + \frac{4}{9} =$
6. $\frac{3}{5} + \frac{2}{3} =$
7. $\frac{5}{9} + \frac{4}{7} =$
8. $\frac{4}{5} + \frac{4}{9} =$
9. $\frac{15}{16} + \frac{5}{6} =$
10. $\frac{7}{12} + \frac{5}{8} =$
11. $\frac{15}{32} + \frac{7}{24} =$
12. $\frac{5}{9} + \frac{9}{16} =$
13. $\frac{1}{2} + \frac{2}{3} + \frac{3}{4} =$
14. $\frac{2}{3} + \frac{5}{8} + \frac{7}{12} =$
15. $\frac{7}{12} + \frac{5}{8} + \frac{5}{8} + \frac{2}{3} =$
16. $\frac{15}{32} + \frac{3}{8} + \frac{7}{16} + \frac{17}{24} =$

1. _____ (10)
2. _____ (10)
3. _____ (10)
4. _____ (10)
5. _____ (10)
6. _____ (10)
7. _____ (12)
8. _____ (12)
9. _____ (13)
10. _____ (13)
11. _____ (13)
12. _____ (13)
13. $\frac{3}{6} = \frac{1}{2}$ (15)
14. $\frac{13}{17}$ (15)
15. $\frac{18}{26} = \frac{9}{13}$ (17)
16. $\frac{39}{64}$ (17)

TEST 41

Name _____ Total Score _____
Date _____ Basic Score 100
Basic Time–7 Minutes   Improvement Score _____

Change to equivalent fractions with common denominators and add. Place each answer in the appropriate space on the right.

1. $\dfrac{3}{4} + \dfrac{1}{6} =$ 2. $\dfrac{3}{8} + \dfrac{5}{6} =$

3. $\dfrac{7}{9} + \dfrac{5}{8} =$ 4. $\dfrac{3}{5} + \dfrac{4}{9} =$

5. $\dfrac{5}{6} + \dfrac{7}{8} =$ 6. $\dfrac{3}{8} + \dfrac{7}{12} =$

7. $\dfrac{15}{16} + \dfrac{5}{9} =$ 8. $\dfrac{8}{25} + \dfrac{1}{2} =$

9. $\dfrac{5}{12} + \dfrac{7}{16} =$ 10. $\dfrac{7}{24} + \dfrac{5}{9} =$

11. $\dfrac{3}{4} + \dfrac{3}{8} + \dfrac{5}{6} =$ 12. $\dfrac{3}{5} + \dfrac{2}{3} + \dfrac{5}{8} =$

13. $\dfrac{2}{3} + \dfrac{1}{2} + \dfrac{4}{5} =$ 14. $\dfrac{7}{8} + \dfrac{2}{3} + \dfrac{1}{6} =$

15. $\dfrac{3}{4} + \dfrac{5}{8} + \dfrac{9}{16} + \dfrac{5}{12} =$

16. $\dfrac{5}{16} + \dfrac{3}{5} + \dfrac{2}{3} + \dfrac{5}{8} =$

1. _____ (10)
2. _____ (10)
3. _____ (10)
4. _____ (10)
5. _____ (10)
6. _____ (10)
7. _____ (11)
8. _____ (11)
9. _____ (12)
10. _____ (12)
11. _____ (15)
12. _____ (15)
13. _____ (15)
14. _____ (15)
15. _____ (17)
16. _____ (17)

42 Addition of Mixed Numbers

To add mixed numbers, (1) add the whole numbers, (2) add the fractions, and (3) reduce the answer to lowest terms.

See the example at the right for the addition of $4\frac{2}{3} + 6\frac{3}{4}$.

$$\text{Add: } 4\frac{2}{3} \qquad \text{Solution: } 4\frac{2}{3} = 4\frac{8}{12}$$
$$\phantom{\text{Add: }} 6\frac{3}{4} \qquad \phantom{\text{Solution: }} 6\frac{3}{4} = 6\frac{9}{12}$$
$$\phantom{\text{Add: }} \phantom{6\frac{3}{4}} \qquad \phantom{\text{Solution: }} \phantom{6\frac{3}{4}} 10\frac{17}{12} = 11\frac{5}{12}$$

EXERCISE 42/Basic Time—10 Minutes
Estimated time to obtain a basic score of 100.

Add the whole-number portions and the fractional parts of each problem separately and then combine the results. Place each answer in the appropriate space on the right.

1. $3\frac{5}{8}$
   $11\frac{13}{16}$

2. $2\frac{3}{4}$
   $5\frac{7}{12}$

3. $9\frac{1}{2}$
   $5\frac{1}{3}$

4. $4\frac{3}{8}$
   $14\frac{1}{6}$

5. $7\frac{3}{8}$
   $5\frac{1}{3}$

6. $2\frac{2}{3}$
   $3\frac{3}{5}$

7. $14\frac{5}{6}$
   $8\frac{2}{9}$

8. $10\frac{4}{9}$
   $14\frac{5}{7}$

9. $15\frac{7}{8}$
   $8\frac{11}{12}$

10. $6\frac{11}{16}$
    $12\frac{13}{24}$

11. $10\frac{5}{16}$
    $5\frac{7}{12}$

12. $7\frac{3}{8}$
    $4\frac{1}{3}$
    $3\frac{4}{5}$

13. $3\frac{1}{2}$
    $4\frac{2}{3}$
    $5\frac{1}{5}$

14. $6\frac{2}{3}$
    $3\frac{1}{8}$
    $5\frac{1}{2}$

1. _____(12)
2. _____(12)
3. _____(12)
4. _____(12)
5. _____(14)
6. _____(14)
7. _____(14)
8. _____(14)
9. _____(15)
10. _____(15)
11. _____(15)
12. _____(17)
13. _____(17)
14. _____(17)

TEST 42

Name _____ Total Score _____

Date _____ Basic Score 100

Basic Time–8 Minutes Improvement Score _____

Add the whole-number portions and the fractional parts of each problem separately and then combine the results. Place each answer in the appropriate space on the right.

1. $4 \frac{3}{4}$
    $7 \frac{1}{2}$

2. $6 \frac{2}{3}$
    $9 \frac{5}{6}$

3. $8 \frac{1}{6}$
    $5 \frac{7}{12}$

4. $3 \frac{1}{2}$
    $2 \frac{1}{3}$

5. $6 \frac{1}{2}$
    $12 \frac{2}{3}$

6. $7 \frac{1}{2}$
    $8 \frac{1}{5}$

7. $8 \frac{3}{4}$
    $12 \frac{5}{6}$

8. $7 \frac{3}{16}$
    $8 \frac{5}{9}$

9. $14 \frac{3}{8}$
    $17 \frac{2}{3}$

10. $75 \frac{2}{5}$
    $14 \frac{7}{8}$

11. $16 \frac{19}{24}$
    $32 \frac{15}{32}$

12. $3 \frac{1}{3}$
    $1 \frac{5}{8}$
    $4 \frac{1}{2}$

13. $3 \frac{2}{3}$
    $4 \frac{7}{9}$
    $5 \frac{1}{8}$

14. $9 \frac{2}{3}$
    $4 \frac{3}{5}$
    $6 \frac{1}{2}$

1. _____ (12)
2. _____ (12)
3. _____ (12)
4. _____ (12)
5. _____ (14)
6. _____ (14)
7. _____ (14)
8. _____ (14)
9. _____ (15)
10. _____ (15)
11. _____ (15)
12. _____ (17)
13. _____ (17)
14. _____ (17)

92 © Copyright South-Western Publishing Co.

LESSON 43

Subtraction of Common Fractions

When common fractions are to be subtracted, (1) change the fractions to equivalent fractions with common denominators; (2) find the difference between the numerators of the new fractions; (3) place the difference over the common denominator; and (4) if necessary, reduce the answer to lowest terms. See these steps in the example.

Subtract: $\frac{9}{10} - \frac{1}{6} =$

Solution: $\frac{27}{30} - \frac{5}{30} = \frac{22}{30} = \frac{11}{15}$

Step 1   Steps 2, 3   Step 4

EXERCISE 43/Basic Time—9 Minutes

Estimated time to obtain a basic score of 100.

Subtract. Place each answer in the appropriate space on the right.

1. $\frac{3}{4} - \frac{2}{3} =$

2. $\frac{2}{3} - \frac{1}{2} =$

3. $\frac{7}{8} - \frac{2}{3} =$

4. $\frac{2}{3} - \frac{3}{8} =$

5. $\frac{4}{5} - \frac{1}{8} =$

6. $\frac{4}{5} - \frac{7}{12} =$

7. $\frac{9}{10} - \frac{2}{3} =$

8. $\frac{1}{3} - \frac{5}{16} =$

9. $\frac{5}{6} - \frac{5}{8} =$

10. $\frac{4}{5} - \frac{3}{4} =$

11. $\frac{15}{16} - \frac{3}{5} =$

12. $\frac{7}{9} - \frac{3}{8} =$

13. $\frac{6}{7} - \frac{5}{9} =$

14. $\frac{9}{16} - \frac{5}{12} =$

15. $\frac{24}{25} - \frac{24}{35} =$

16. $\frac{15}{18} - \frac{19}{24} =$

1. _____(10)
2. _____(10)
3. _____(10)
4. _____(10)
5. _____(10)
6. _____(10)
7. _____(10)
8. _____(10)
9. _____(10)
10. _____(10)
11. _____(15)
12. _____(15)
13. _____(15)
14. _____(15)
15. _____(20)
16. _____(20)

© Copyright South-Western Publishing Co.

TEST 43

Name _____   Total Score _____
Date _____   Basic Score 100
Basic Time–8 Minutes   Improvement Score _____

Subtract. Place each answer in the appropriate space on the right.

1. $\frac{1}{2} - \frac{1}{3} =$ 2. $\frac{2}{3} - \frac{1}{4} =$

3. $\frac{3}{4} - \frac{1}{6} =$ 4. $\frac{5}{6} - \frac{3}{5} =$

5. $\frac{5}{8} - \frac{3}{7} =$ 6. $\frac{6}{7} - \frac{2}{5} =$

7. $\frac{7}{8} - \frac{5}{6} =$ 8. $\frac{4}{7} - \frac{1}{8} =$

9. $\frac{5}{6} - \frac{5}{8} =$ 10. $\frac{7}{12} - \frac{3}{8} =$

11. $\frac{5}{6} - \frac{5}{16} =$ 12. $\frac{7}{8} - \frac{2}{3} =$

13. $\frac{6}{7} - \frac{4}{5} =$ 14. $\frac{8}{9} - \frac{5}{12} =$

15. $\frac{9}{10} - \frac{1}{6} =$ 16. $\frac{15}{16} - \frac{2}{3} =$

1. _____ (10)
2. _____ (10)
3. _____ (10)
4. _____ (10)
5. _____ (12)
6. _____ (12)
7. _____ (12)
8. _____ (12)
9. _____ (12)
10. _____ (12)
11. _____ (14)
12. _____ (14)
13. _____ (15)
14. _____ (15)
15. _____ (15)
16. _____ (15)

LESSON 44

Subtraction of Mixed Numbers

When mixed numbers are to be subtracted, (1) change each mixed number to an improper fraction; (2) convert the improper fractions to equivalent fractions with common denominators; (3) subtract the numerators; (4) place the difference over the common denominator; and (5) if necessary, reduce to lowest terms or a mixed number. See these steps in the example.

Subtract: $5\frac{1}{3} - 3\frac{1}{2} =$

Solution: $\frac{16}{3} - \frac{7}{2} = \frac{32}{6} - \frac{21}{6} = \frac{11}{6} = 1\frac{5}{6}$

         Step 1   Step 2   Steps 3, 4 Step 5

EXERCISE 44/Basic Time—12 Minutes

Estimated time to obtain a basic score of 100.

Subtract. Place each answer in the appropriate space on the right.

1. $5\frac{3}{4} - 3\frac{1}{4} =$
2. $12\frac{2}{3} - 7\frac{1}{3} =$
3. $3\frac{1}{2} - 1\frac{1}{4} =$
4. $3\frac{3}{8} - 2\frac{1}{2} =$
5. $7\frac{3}{16} - 5\frac{5}{8} =$
6. $7\frac{2}{3} - 3\frac{4}{5} =$
7. $6\frac{3}{4} - 5\frac{5}{9} =$
8. $4\frac{2}{3} - 3\frac{1}{2} =$
9. $7\frac{3}{4} - 5\frac{1}{6} =$
10. $9\frac{3}{5} - 7\frac{1}{2} =$
11. $12\frac{1}{8} - 11\frac{7}{12} =$
12. $13\frac{4}{5} - 9\frac{5}{6} =$
13. $32\frac{1}{8} - 25\frac{1}{12} =$
14. $14\frac{2}{3} - 8\frac{7}{8} =$
15. $35\frac{4}{9} - 18\frac{1}{6} =$
16. $62\frac{5}{12} - 6\frac{3}{5} =$

1. _____(10)
2. _____(10)
3. _____(10)
4. _____(10)
5. _____(12)
6. _____(12)
7. _____(12)
8. _____(12)
9. _____(12)
10. _____(12)
11. _____(14)
12. _____(14)
13. _____(15)
14. _____(15)
15. _____(15)
16. _____(15)

© Copyright South-Western Publishing Co.

TEST 44

Name _____ Total Score _____
Date _____ Basic Score __100__
Basic Time–10 Minutes Improvement Score _____

Subtract. Place each answer in the appropriate space on the right.

1. $7\frac{1}{2} - 1\frac{1}{4} =$ 2. $6\frac{2}{3} - 3\frac{5}{6} =$

3. $10\frac{7}{8} - 6\frac{3}{4} =$ 4. $9\frac{11}{16} - 7\frac{3}{4} =$

5. $4\frac{7}{8} - 2\frac{3}{5} =$ 6. $6\frac{5}{8} - 4\frac{1}{6} =$

7. $5\frac{1}{6} - 2\frac{3}{4} =$ 8. $9\frac{4}{5} - 8\frac{2}{3} =$

9. $7\frac{2}{5} - 6\frac{1}{4} =$ 10. $16\frac{1}{3} - 7\frac{7}{8} =$

11. $3\frac{1}{3} - 2\frac{1}{2} =$ 12. $6\frac{2}{3} - 5\frac{5}{8} =$

13. $9\frac{2}{3} - 8\frac{4}{5} =$ 14. $12\frac{3}{4} - 5\frac{5}{6} =$

15. $12\frac{1}{4} - 8\frac{2}{3} =$ 16. $15\frac{1}{2} - 12\frac{1}{8} =$

1. _____ (10)
2. _____ (10)
3. _____ (10)
4. _____ (10)
5. _____ (12)
6. _____ (12)
7. _____ (12)
8. _____ (12)
9. _____ (12)
10. _____ (12)
11. _____ (14)
12. _____ (14)
13. _____ (15)
14. _____ (15)
15. _____ (15)
16. _____ (15)

LESSON 45: Multiplication of Common Fractions

To find a fractional part *of* a number, multiply the number by the fraction. To multiply common fractions, see Example A.

Example A $\frac{7}{8} \times \frac{5}{6} = \frac{7 \times 5}{8 \times 6} = \frac{35}{48}$

Solution A Multiply the numerators, giving $7 \times 5 = 35$, the numerator of the result. Multiply the denominators, giving $8 \times 6 = 48$, the denominator of the result.

To divide the numerators and denominators by common divisors before multiplying, see Example B.

Example B $\frac{3}{8} \times \frac{5}{6} \times \frac{4}{5} = \frac{\overset{1}{3}}{\underset{2}{8}} \times \frac{\overset{1}{5}}{\underset{2}{6}} \times \frac{\overset{1}{4}}{\underset{1}{5}} = \frac{1}{4}$

Solution B Divide any numerator and denominator by a divisor that is common to both. In Example B the first numerator, 3, and the second denominator, 6, have a common divisor of 3. Dividing both by 3 gives 1 and 2, respectively. The third numerator, 4, and the first denominator, 8, may both be divided by 4, giving 1 and 2, respectively. The second numerator, 5, and the third denominator, 5, may both be divided by 5, giving a 1 in each place. Multiplying the new numerators ($1 \times 1 \times 1$) gives 1 as the numerator of the answer. Multiplying the new denominators ($2 \times 2 \times 1$) produces 4 as the denominator of the answer. The product of the fractions, then, is $\frac{1}{4}$. When the numerators and denominators have no common divisors, proceed as in Example A.

EXERCISE 45/Basic Time—5 Minutes

Estimated time to obtain a basic score of 100.

Multiply:

1. $\frac{3}{8} \times \frac{3}{5} =$ 2. $\frac{5}{9} \times \frac{7}{12} =$ 3. $\frac{4}{5} \times \frac{2}{3} =$ (6)

4. $\frac{5}{8} \times \frac{7}{8} =$ 5. $\frac{2}{3} \times \frac{16}{25} =$ 6. $\frac{2}{9} \times \frac{7}{11} =$ (6)

7. $\frac{15}{16} \times \frac{3}{4} =$ 8. $\frac{9}{10} \times \frac{7}{8} =$ 9. $\frac{11}{12} \times \frac{5}{16} =$ (6)

10. $\frac{19}{24} \times \frac{5}{6} =$ 11. $\frac{4}{5} \times \frac{3}{16} =$ 12. $\frac{7}{12} \times \frac{5}{8} =$ (6)

13. $\frac{7}{8} \times \frac{6}{7} =$ 14. $\frac{4}{9} \times \frac{9}{10} =$ 15. $\frac{4}{5} \times \frac{5}{8} =$ (6)

16. $\frac{15}{16} \times \frac{32}{45} =$ 17. $\frac{7}{16} \times \frac{4}{5} =$ 18. $\frac{3}{16} \times \frac{4}{15} =$ (6)

19. $\frac{3}{8} \times \frac{9}{10} \times \frac{5}{9} \times \frac{2}{3} =$ 20. $\frac{9}{10} \times \frac{5}{12} \times \frac{16}{21} \times \frac{3}{4} =$ (10)

21. $\frac{3}{16} \times \frac{4}{5} \times \frac{2}{3} \times \frac{3}{8} =$ 22. $\frac{7}{8} \times \frac{2}{3} \times \frac{1}{2} \times \frac{5}{7} =$ (10)

23. $\frac{5}{12} \times \frac{7}{16} \times \frac{8}{15} \times \frac{16}{21} =$ 24. $\frac{4}{5} \times \frac{5}{7} \times \frac{7}{8} \times \frac{1}{2} =$ (12)

25. $\frac{13}{15} \times \frac{5}{12} \times \frac{8}{9} \times \frac{3}{4} =$ 26. $\frac{7}{9} \times \frac{5}{12} \times \frac{3}{14} \times \frac{3}{5} =$ (14)

TEST 45

Name _____ Total Score _____
Date _____ Basic Score __100__
Basic Time–2½ Minutes Improvement Score _____

Multiply. Place each answer in the appropriate space on the right.

1. $\dfrac{5}{6} \times \dfrac{7}{8} =$ 2. $\dfrac{3}{5} \times \dfrac{4}{5} =$

3. $\dfrac{7}{16} \times \dfrac{5}{12} =$ 4. $\dfrac{4}{5} \times \dfrac{5}{8} =$

5. $\dfrac{3}{4} \times \dfrac{5}{12} =$ 6. $\dfrac{3}{8} \times \dfrac{6}{7} =$

7. $\dfrac{2}{3} \times \dfrac{7}{8} =$ 8. $\dfrac{9}{16} \times \dfrac{8}{9} =$

9. $\dfrac{4}{9} \times \dfrac{5}{8} =$ 10. $\dfrac{7}{15} \times \dfrac{24}{35} =$

11. $\dfrac{2}{3} \times \dfrac{4}{5} \times \dfrac{5}{12} =$

12. $\dfrac{3}{8} \times \dfrac{11}{12} \times \dfrac{16}{33} =$

13. $\dfrac{7}{8} \times \dfrac{3}{5} \times \dfrac{5}{14} \times \dfrac{4}{5} \times \dfrac{2}{3} =$

14. $\dfrac{7}{16} \times \dfrac{8}{9} \times \dfrac{2}{7} \times \dfrac{5}{6} \times \dfrac{1}{2} =$

15. $\dfrac{1}{5} \times \dfrac{2}{3} \times \dfrac{5}{12} \times \dfrac{6}{7} \times \dfrac{7}{8} =$

1. _____ (8)
2. _____ (8)
3. _____ (8)
4. _____ (8)
5. _____ (8)
6. _____ (8)
7. _____ (8)
8. _____ (8)
9. _____ (8)
10. _____ (12)
11. _____ (12)
12. _____ (18)
13. _____ (28)
14. _____ (30)
15. _____ (28)

© Copyright South-Western Publishing Co.

LESSON 46

Multiplication of Mixed Numbers

To multiply mixed numbers, change each mixed number to its equivalent improper fraction. Then multiply in the manner described in Exercise 45. If possible, use cancellation by dividing the numerators and denominators by common divisors before proceeding with the multiplication. Canceling first gives smaller numerators and denominators to multiply. See the example.

Multiply

$$1\frac{2}{5} \times 3\frac{1}{3} \times 1\frac{1}{7} = \frac{\overset{1}{7}}{5} \times \frac{\overset{2}{10}}{3} \times \frac{8}{\underset{1}{7}} = \frac{16}{3} = 5\frac{1}{3}$$

EXERCISE 46/Basic Time – 10 Minutes
Estimated time to obtain a basic score of 100.

Multiply. Place each answer in the appropriate space on the right.

1. $3\frac{1}{2} \times 4\frac{2}{3} =$

2. $5\frac{1}{4} \times 2\frac{2}{3} =$

3. $7\frac{3}{8} \times 9\frac{5}{6} =$

4. $8\frac{5}{9} \times 4\frac{1}{11} =$

5. $12\frac{3}{5} \times 10\frac{5}{9} =$

6. $6\frac{5}{8} \times 12\frac{6}{7} =$

7. $17\frac{1}{2} \times 19\frac{2}{5} =$

8. $14\frac{2}{7} \times 16\frac{3}{5} =$

9. $23\frac{4}{7} \times 19\frac{5}{8} =$

10. $35\frac{7}{16} \times 24\frac{5}{12} =$

11. $28\frac{11}{12} \times 15\frac{5}{6} =$

12. $18\frac{8}{9} \times 26\frac{5}{7} =$

1. _____ (15)
2. _____ (15)
3. _____ (15)
4. _____ (15)
5. _____ (15)
6. _____ (15)
7. _____ (15)
8. _____ (15)
9. _____ (20)
10. _____ (20)
11. _____ (20)
12. _____ (20)

TEST 46

Name _____ Total Score _____

Date _____ Basic Score 100

Basic Time–10 Minutes Improvement Score _____

Multiply. Place each answer in the appropriate space on the right.

1. $4\frac{2}{3} \times 6\frac{1}{2} =$

2. $9\frac{3}{4} \times 10\frac{1}{3} =$

3. $5\frac{3}{8} \times 7\frac{3}{5} =$

4. $8\frac{5}{6} \times 3\frac{2}{7} =$

5. $11\frac{5}{9} \times 6\frac{1}{12} =$

6. $6\frac{7}{8} \times 4\frac{15}{16} =$

7. $12\frac{4}{7} \times 9\frac{6}{11} =$

8. $15\frac{1}{3} \times 16\frac{2}{3} =$

9. $25\frac{3}{5} \times 16\frac{7}{8} =$

10. $16\frac{1}{8} \times 18\frac{5}{9} =$

11. $3\frac{24}{25} \times 5\frac{15}{16} =$

12. $8\frac{11}{12} \times 12\frac{7}{11} =$

1. _____ (15)
2. _____ (15)
3. _____ (15)
4. _____ (15)
5. _____ (15)
6. _____ (15)
7. _____ (15)
8. _____ (15)
9. _____ (20)
10. _____ (20)
11. _____ (20)
12. _____ (20)

LESSON 47
Multiplication of Whole Numbers by Common Fractions

Any problem requiring multiplication of whole numbers by a common fraction may be interpreted to mean:

$$\frac{\text{Numerator} \times \text{Numerator}}{\text{Denominator} \times \text{Denominator}} = \text{Product}$$

In Example A below, the number 37 means 37 ones and may be written as $\frac{37}{1}$. Therefore, one method for solving the problem is to multiply numerator times numerator (37 × 1) and denominator times denominator (1 × 5), which gives $\frac{37}{5}$. Reducing this improper fraction to lowest terms reveals the product to be $7\frac{2}{5}$.

Example A $37 \times \frac{1}{5} = \frac{37}{1} \times \frac{1}{5} = \frac{37}{5} = 7\frac{2}{5}$

Canceling when possible is recommended. Notice that canceling the 1's in Example A would give the same improper fraction. As Example B shows, canceling first gives smaller numbers to multiply.

Example B $45 \times \frac{2}{3} = \frac{\overset{15}{45}}{1} \times \frac{2}{\underset{1}{3}} = \frac{30}{1} = 30$

If you cannot cancel easily, multiply the numerators. Then divide the result by the product of the denominators, and reduce to lowest terms. See Example C.

Example C $57 \times \frac{5}{6} = \frac{57}{1} \times \frac{5}{6} = \frac{285}{6} = 47\frac{3}{6} = 47\frac{1}{2}$

EXERCISE 47/Basic Time—9 Minutes
Estimated time to obtain a basic score of 100.

Multiply. Show each answer in lowest terms. Place each answer in the appropriate space on the right.

1. $130 \times \frac{1}{5} =$
2. $268 \times \frac{1}{4} =$
3. $174 \times \frac{1}{2} =$
4. $135 \times \frac{1}{3} =$
5. $168 \times \frac{1}{7} =$
6. $232 \times \frac{1}{8} =$
7. $279 \times \frac{1}{6} =$
8. $472 \times \frac{1}{9} =$
9. $\frac{1}{8} \times 387 =$
10. $\frac{1}{5} \times 367 =$
11. $\frac{3}{4} \times 72 =$
12. $186 \times \frac{2}{3} =$
13. $\frac{5}{6} \times 243 =$
14. $590 \times \frac{7}{8} =$
15. $\frac{12}{13} \times 58 =$
16. $57 \times \frac{15}{17} =$

1. _____ (10)
2. _____ (10)
3. _____ (10)
4. _____ (10)
5. _____ (10)
6. _____ (10)
7. _____ (10)
8. _____ (10)
9. _____ (10)
10. _____ (10)
11. _____ (15)
12. _____ (15)
13. _____ (15)
14. _____ (15)
15. _____ (20)
16. _____ (20)

TEST 47

Name _____ Total Score _____
Date _____ Basic Score __100__
Basic Time—9 Minutes Improvement Score _____

Multiply. Show each answer in lowest terms. Place each answer in the appropriate space on the right.

1. $48 \times \frac{1}{3} =$

2. $346 \times \frac{1}{2} =$

3. $\frac{1}{5} \times 245 =$

4. $308 \times \frac{1}{4} =$

5. $1{,}026 \times \frac{1}{6} =$

6. $\frac{1}{7} \times 525 =$

7. $\frac{1}{9} \times 172 =$

8. $664 \times \frac{1}{8} =$

9. $\frac{1}{7} \times 342 =$

10. $99 \times \frac{1}{6} =$

11. $810 \times \frac{2}{3} =$

12. $\frac{3}{4} \times 723 =$

13. $\frac{5}{7} \times 244 =$

14. $325 \times \frac{7}{8} =$

15. $172 \times \frac{11}{12} =$

16. $138 \times \frac{13}{15} =$

1. _____ (10)
2. _____ (10)
3. _____ (10)
4. _____ (10)
5. _____ (10)
6. _____ (10)
7. _____ (10)
8. _____ (10)
9. _____ (10)
10. _____ (10)
11. _____ (15)
12. _____ (15)
13. _____ (15)
14. _____ (15)
15. _____ (20)
16. _____ (20)

© Copyright South-Western Publishing Co.

LESSON 48: Division of Common Fractions

To divide one common fraction by another, invert the divisor (that is, turn it upside down) and then multiply the fractions as described in Exercise 45. Reduce the quotient to lowest terms or a mixed number, if necessary.

See the example at the right, where $\frac{3}{4}$ is divided by $\frac{2}{3}$. Note that the divisor $\frac{2}{3}$ is inverted to $\frac{3}{2}$ and the division sign is changed to a multiplication sign. The two fractions are then multiplied, and the quotient reduced to a mixed number. The fraction $\frac{3}{4}$ is $1\frac{1}{8}$ times as large as the fraction $\frac{2}{3}$.

Divide: $\frac{3}{4} \div \frac{2}{3} = \frac{3}{4} \times \frac{3}{2} = \frac{9}{8} = 1\frac{1}{8}$

EXERCISE 48/Basic Time–3 Minutes

Estimated time to obtain a basic score of 100.

Divide. Place each answer in the appropriate space on the right.

1. $\frac{5}{8} \div \frac{2}{7} =$

2. $\frac{5}{12} \div \frac{7}{8} =$

3. $\frac{7}{15} \div \frac{4}{9} =$

4. $\frac{5}{16} \div \frac{1}{8} =$

5. $\frac{5}{9} \div \frac{7}{8} =$

6. $\frac{9}{16} \div \frac{7}{12} =$

7. $\frac{11}{12} \div \frac{14}{15} =$

8. $\frac{13}{18} \div \frac{8}{9} =$

9. $\frac{19}{32} \div \frac{17}{24} =$

10. $\frac{22}{27} \div \frac{28}{33} =$

11. $\frac{25}{42} \div \frac{5}{21} =$

12. $\frac{49}{72} \div \frac{21}{32} =$

13. $\frac{51}{56} \div \frac{17}{24} =$

14. $\frac{16}{25} \div \frac{36}{55} =$

1. _____(10)
2. _____(10)
3. _____(10)
4. _____(10)
5. _____(12)
6. _____(12)
7. _____(12)
8. _____(12)
9. _____(18)
10. _____(18)
11. _____(18)
12. _____(18)
13. _____(20)
14. _____(20)

TEST 48

Name _____ Total Score _____
Date _____ Basic Score __100__
Basic Time–3 Minutes Improvement Score _____

Divide. Place each answer in the appropriate space on the right.

1. $\dfrac{4}{5} \div \dfrac{7}{8} =$

2. $\dfrac{5}{16} \div \dfrac{4}{15} =$

3. $\dfrac{7}{12} \div \dfrac{15}{16} =$

4. $\dfrac{5}{9} \div \dfrac{7}{11} =$

5. $\dfrac{9}{16} \div \dfrac{5}{12} =$

6. $\dfrac{17}{24} \div \dfrac{19}{32} =$

7. $\dfrac{24}{27} \div \dfrac{8}{9} =$

8. $\dfrac{27}{32} \div \dfrac{9}{16} =$

9. $\dfrac{24}{25} \div \dfrac{32}{45} =$

10. $\dfrac{62}{63} \div \dfrac{31}{42} =$

11. $\dfrac{18}{25} \div \dfrac{5}{6} =$

12. $\dfrac{25}{27} \div \dfrac{17}{18} =$

13. $\dfrac{13}{22} \div \dfrac{23}{25} =$

14. $\dfrac{29}{36} \div \dfrac{18}{19} =$

1. _____ (10)
2. _____ (10)
3. _____ (10)
4. _____ (10)
5. _____ (12)
6. _____ (12)
7. _____ (12)
8. _____ (12)
9. _____ (18)
10. _____ (18)
11. _____ (18)
12. _____ (18)
13. _____ (20)
14. _____ (20)

104 © Copyright South-Western Publishing Co.

LESSON 49 — Division of Mixed Numbers

To divide one mixed number by another, change each mixed number to an improper fraction. Then proceed as in the division of common fractions described in Exercise 48.

Divide: $3\frac{2}{3} \div 2\frac{1}{2} = \frac{11}{3} \div \frac{5}{2} = \frac{11}{3} \times \frac{2}{5} = \frac{22}{15} = 1\frac{7}{15}$

EXERCISE 49/Basic Time—7 Minutes

Estimated time to obtain a basic score of 100.

Divide. Place each answer in the appropriate space on the right.

1. $5\frac{1}{2} \div 3\frac{1}{3} =$ 2. $6\frac{3}{4} \div 5\frac{1}{2} =$

3. $3\frac{3}{5} \div 5\frac{1}{4} =$ 4. $7\frac{1}{3} \div 2\frac{3}{4} =$

5. $8\frac{5}{6} \div 6\frac{2}{3} =$ 6. $2\frac{3}{8} \div 4\frac{1}{6} =$

7. $13\frac{5}{8} \div 7\frac{1}{8} =$ 8. $12\frac{6}{7} \div 15\frac{4}{5} =$

9. $32\frac{11}{35} \div 19\frac{17}{20} =$

10. $24\frac{9}{10} \div 7\frac{15}{44} =$

11. $64\frac{1}{3} \div 24\frac{2}{3} =$

12. $16\frac{1}{4} \div 8\frac{1}{3} =$

1. _____(12)
2. _____(12)
3. _____(12)
4. _____(12)
5. _____(16)
6. _____(16)
7. _____(16)
8. _____(16)
9. _____(22)
10. _____(22)
11. _____(22)
12. _____(22)

© Copyright South-Western Publishing Co.

Name _____ Total Score _____
Date _____ Basic Score 100
Basic Time–5 Minutes   Improvement Score _____

Divide. Place each answer in the appropriate space on the right.

1. $2\frac{3}{8} \div 1\frac{3}{4} =$ 2. $7\frac{1}{2} \div 5\frac{2}{3} =$

3. $9\frac{3}{5} \div 6\frac{3}{4} =$ 4. $12\frac{2}{5} \div 9\frac{5}{6} =$

5. $10\frac{5}{8} \div 5\frac{5}{6} =$ 6. $8\frac{4}{5} \div 16\frac{2}{3} =$

7. $25\frac{5}{9} \div 12\frac{5}{6} =$ 8. $32\frac{3}{5} \div 21\frac{3}{4} =$

9. $19\frac{13}{16} \div 22\frac{7}{8} =$ 10. $24\frac{19}{25} \div 16\frac{47}{75} =$

11. $16\frac{3}{8} \div 17\frac{7}{12} =$ 12. $31\frac{5}{9} \div 25\frac{2}{3} =$

13. $27\frac{6}{7} \div 9\frac{2}{3} =$ 14. $14\frac{2}{7} \div 24\frac{9}{14} =$

1. _____(10)
2. _____(10)
3. _____(10)
4. _____(10)
5. _____(12)
6. _____(12)
7. _____(12)
8. _____(12)
9. _____(18)
10. _____(18)
11. _____(18)
12. _____(18)
13. _____(20)
14. _____(20)

LESSON 50
Simplification of Complex Fractions

A fraction that has a fraction or a mixed number, or both, for its numerator or denominator is a *complex fraction*. In a common fraction, both the numerator and denominator (dividend and divisor) are whole numbers. The line between the numerator and the denominator of a fraction, whether common or complex, is simply one way of showing division. When a complex fraction has been reduced to lowest terms, it is said to have been *simplified*.

To simplify a complex fraction, perform the division indicated, following the rules for division of common fractions. Remember that 1 is understood to be the denominator of a whole number.

Simplify: $\dfrac{5\frac{1}{4}}{6} = 5\frac{1}{4} \div 6 = \dfrac{\overset{7}{\cancel{21}}}{4} \times \dfrac{1}{\underset{2}{\cancel{6}}} = \dfrac{7}{8}$

EXERCISE 50/Basic Time–6 Minutes
Estimated time to obtain a basic score of 100.

Simplify these complex fractions. Place each answer in the appropriate space on the right. (20 each)

1. $\dfrac{5}{\frac{3}{4}} =$

2. $\dfrac{6\frac{1}{2}}{8\frac{1}{4}} =$

3. $\dfrac{8\frac{1}{3}}{100} =$

4. $\dfrac{\frac{5}{6}}{7\frac{1}{2}} =$

5. $\dfrac{\frac{2}{3}}{\frac{5}{6}} =$

6. $\dfrac{3\frac{2}{3}}{8\frac{3}{4}} =$

7. $\dfrac{\frac{1}{4}}{\frac{7}{8}} =$

8. $\dfrac{10\frac{4}{5}}{12\frac{5}{8}} =$

9. $\dfrac{\frac{3}{5}}{\frac{3}{7}} =$

10. $\dfrac{87\frac{1}{2}}{100} =$

1. _____
2. _____
3. _____
4. _____
5. _____
6. _____
7. _____
8. _____
9. _____
10. _____

© Copyright South-Western Publishing Co.

TEST 50

Name _____   Total Score _____
Date _____   Basic Score 100
Basic Time–5 Minutes   Improvement Score _____

Simplify these complex fractions. Place each answer in the appropriate space on the right. (20 each)

1. $\dfrac{12\frac{1}{2}}{100} =$ 2. $\dfrac{12\frac{1}{5}}{7\frac{1}{2}} =$

3. $\dfrac{\frac{1}{9}}{10} =$ 4. $\dfrac{9\frac{5}{12}}{12\frac{3}{4}} =$

5. $\dfrac{3\frac{1}{3}}{10} =$ 6. $\dfrac{7\frac{2}{3}}{24\frac{5}{9}} =$

7. $\dfrac{9}{\frac{3}{4}} =$ 8. $\dfrac{5\frac{1}{4}}{16\frac{4}{5}} =$

9. $\dfrac{\frac{5}{8}}{\frac{5}{12}} =$ 10. $\dfrac{27\frac{1}{3}}{36\frac{3}{4}} =$

1. _____
2. _____
3. _____
4. _____
5. _____
6. _____
7. _____
8. _____
9. _____
10. _____

LESSON 51
Review of Fractions

This lesson is a review of fractions. Whenever possible, use one of the short methods presented in the preceding exercises. In any case, use the method that gives you the greatest accuracy.

EXERCISE 51/Basic Time–7 Minutes
Estimated time to obtain a basic score of 100.

Solve these problems. Place each answer in the appropriate space on the right.

Change to lowest terms:

1. $\frac{64}{160} =$
2. $\frac{90}{135} =$

Add:

3. $\frac{3}{4} + \frac{5}{6} + \frac{7}{8} =$
4. $26\frac{2}{3} + 12\frac{5}{16} =$

Subtract:

5. $\frac{11}{12} - \frac{7}{8} =$
6. $15\frac{2}{3} - 12\frac{3}{4} =$

Multiply:

7. $\frac{3}{4} \times \frac{5}{16} \times \frac{1}{2} =$
8. $\frac{9}{16} \times \frac{2}{3} \times \frac{7}{8} \times \frac{9}{14} =$

9. $18\frac{3}{4} \times 6\frac{2}{3} =$
10. $25\frac{5}{12} \times 15\frac{3}{8} =$

Divide:

11. $\frac{3}{8} \div \frac{5}{12} =$
12. $\frac{11}{16} \div \frac{7}{12} =$

13. $6\frac{2}{3} \div 12\frac{1}{2} =$
14. $18\frac{3}{4} \div 5\frac{5}{6} =$

1. _____(10)
2. _____(10)
3. _____(10)
4. _____(10)
5. _____(12)
6. _____(12)
7. _____(12)
8. _____(12)
9. _____(18)
10. _____(18)
11. _____(18)
12. _____(18)
13. _____(20)
14. _____(20)

© Copyright South-Western Publishing Co.

TEST 51

Name _____ Total Score _____
Date _____ Basic Score __100__
Basic Time–9 Minutes Improvement Score _____

Solve these problems. Place each answer in the appropriate space on the right.

Add:

1. $\frac{1}{2} + \frac{1}{3} + \frac{1}{4} + \frac{1}{6} =$

2. $\frac{3}{4} + \frac{5}{6} + \frac{7}{8} + \frac{5}{12} =$

3. $16\frac{2}{3} + \frac{7}{16} + 9\frac{3}{4} =$

4. $9\frac{5}{6} + 7\frac{7}{16} + 3\frac{11}{18} =$

Subtract:

5. $\frac{5}{12} - \frac{3}{16} =$

6. $\frac{15}{16} - \frac{11}{24} =$

7. $17\frac{7}{8} - 9\frac{1}{3} =$

8. $12\frac{1}{2} - 8\frac{8}{9} =$

Multiply:

9. $\frac{15}{16} \times \frac{23}{30} =$

10. $\frac{3}{4} \times \frac{5}{6} \times \frac{7}{8} =$

11. $16\frac{2}{3} \times 12\frac{1}{2} =$

12. $7\frac{11}{12} \times 6\frac{9}{16} =$

Divide:

13. $\frac{63}{64} \div \frac{9}{16} =$

14. $35\frac{7}{12} \div 16\frac{2}{3} =$

1. _____ (10)
2. _____ (10)
3. _____ (10)
4. _____ (10)
5. _____ (12)
6. _____ (12)
7. _____ (12)
8. _____ (12)
9. _____ (18)
10. _____ (18)
11. _____ (18)
12. _____ (18)
13. _____ (20)
14. _____ (20)

LESSON 52
Application Problems

EXERCISE 52/Basic Time—8 Minutes

Estimated time to obtain a basic score of 100.

Solve these problems. Place each answer in the appropriate space on the right. (20 each)

1. Phil weighs $226\frac{5}{6}$ pounds when wearing all of his scuba gear, which weighs $54\frac{1}{2}$ pounds. How much does Phil weigh without the scuba gear?

    1._____

2. Which problem has the larger sum: (a) $\frac{3}{8}+\frac{2}{9}+\frac{5}{6}$ or (b) $\frac{1}{9}+\frac{3}{5}+\frac{7}{10}$? How much larger?

    2._____

3. A homeowner plans to place some 12-inch boards on shelf brackets in the closet to make 3 shelves $4\frac{3}{4}$ feet long and 4 shelves $3\frac{1}{3}$ feet long. How many feet of 12-inch boards are needed for these shelves?

    3._____

4. A farmer harvested 56 of the 72 acres that were planted in corn. What fractional part of the corn crop was harvested?

    4._____

5. From the cabin to the lake is $\frac{3}{4}$ mile. Four people each carried the boat oars $\frac{1}{4}$ of the way to the lake. What fractional part of a mile did each person carry the oars?

    5._____

6. Last year Jeanne's salary was $24,000. This year her salary is $\frac{3}{20}$ more. How much is her salary this year?

    6._____

7. A caterer who has 4 large cakes wishes to serve 64 equal portions. Each portion will be what fractional part of a cake?

    7._____

8. Don had a difficult time rowing up a swift stream. He rowed at a rate of $\frac{3}{5}$ mile an hour. At this rate, how long did it take him to row $3\frac{1}{3}$ miles?

    8._____

9. Diane drove 18 miles from her house. This was $\frac{2}{3}$ of the distance she had to drive in order to reach the theater. How many miles did she have to drive from her house to reach the theater?

    9._____

10. How many pies are needed to serve $\frac{1}{6}$ pie to each of 54 guests?

    10._____

Name _____ Total Score _____
Date _____ Basic Score __100__
Basic Time–7 Minutes   Improvement Score _____

Solve these problems. Write the answers in the spaces on the right. (20 each)

1. A company's stock sold at $95\frac{3}{8}$ per share when the stock market opened for the day. The stock sold for $92\frac{5}{8}$ when the stock market closed that day. How many points per share did this stock drop that day?   1. _____

2. To expand its runways, an airport bought five tracts of land: $62\frac{1}{3}$ acres, $14\frac{4}{5}$ acres, $33\frac{2}{3}$ acres, $49\frac{7}{15}$ acres, and $10\frac{3}{5}$ acres. How many acres did the airport purchase?   2. _____

3. Norman bought half of a watermelon. He shared it with three friends. If Norman divided the watermelon equally, what part of a whole watermelon did each person receive?   3. _____

4. The Garcia family spent $\frac{1}{5}$ of its income for rent, $\frac{1}{4}$ for food, $\frac{1}{3}$ for clothing, $\frac{1}{6}$ for miscellaneous items, and saved the remainder. What fractional part of the income was saved?   4. _____

5. A delivery truck averages $13\frac{1}{2}$ miles per gallon of gasoline. How many gallons of gasoline were used to drive this truck $1{,}098\frac{9}{10}$ miles?   5. _____

6. A pilot flew a plane 555 miles per hour for $1\frac{5}{6}$ hours. How many miles did she fly?   6. _____

7. How many pieces of fabric $3\frac{1}{3}$ yards long can be cut from a bolt of fabric that is 40 yards long?   7. _____

8. A farmer sold $62\frac{1}{2}$ acres of a 750-acre farm. Expressed as a fraction, what part of the farm was sold?   8. _____

9. A water tank held $719\frac{3}{4}$ gallons. If 17 barrels, each containing $37\frac{1}{2}$ gallons, were drawn from it, how many gallons would remain in the tank?   9. _____

10. The partnership of Lee, Cohen, and O'Hara agreed to share all gains and losses in proportion to their respective investments. Lee invested $21,000; Cohen, $35,000; and O'Hara, $28,000. If they are to share a net gain of $21,600, how much should (a) Lee, (b) Cohen, and (c) O'Hara receive?

10a. _____ (6)
10b. _____ (7)
10c. _____ (7)

3 Cumulative Review

Name _____

Complete each statement by supplying the missing word(s).

1. A proper fraction is one in which the numerator is (equal to/smaller than) ? the denominator.

    1._____

2. An improper fraction is one in which the numerator is not (smaller/larger) ? than the denominator.

    2._____

3. Multiplying both the numerator and denominator of a fraction by the same number, except 0, gives a fraction of (equal/greater) ? value.

    3._____

4. Dividing both the numerator and denominator of a fraction by the same number, except 0, gives a fraction of (equal/smaller) ? value.

    4._____

5. When both the numerator and denominator of a fraction are divided by the same number, except 0, an equivalent fraction in (smaller/larger) ? terms is obtained.

    5._____

6. When the numerator and denominator of a fraction have no common divisor except 1, the fraction is in (lowest/greatest) ? terms.

    6._____

7. The lowest number into which both the numerator and denominator will divide without a remainder is called the ? .

    7._____

8. The largest number that will divide into both the denominator and the numerator of a fraction is called the ? .

    8._____

9. Any number larger than 1 that is divisible only by 1 and by itself is called a ? number.

    9._____

10. The number 1 equals ? - ninths.

    10._____

11. Unlike fractions (should/should not) ? be changed to fractions with a common denominator before being added or subtracted.

    11._____

Change each fraction to the higher or lower terms shown.

12. $\dfrac{2}{7} = \dfrac{?}{224}$

    12._____

13. $\dfrac{80}{128} = \dfrac{?}{8}$

    13._____

14. $\dfrac{7}{12} = \dfrac{?}{252}$

    14._____

Find the CGD and reduce each fraction to lowest terms.

15. GCD ? ; $\dfrac{65}{156}$

    15._____

16. GCD ? ; $\dfrac{57}{95}$

    16._____

3 Cumulative Review

Change each improper fraction to a mixed number in lowest terms.

17. $\dfrac{235}{8} =$

18. $\dfrac{88}{5} =$

19. $\dfrac{274}{7} =$

Change each mixed number to an improper fraction.

20. $23\dfrac{7}{8} =$

21. $19\dfrac{2}{3} =$

22. $70\dfrac{7}{9} =$

Find the lowest common denominator for each set of fractions.

23. $\dfrac{5}{9}, \dfrac{7}{12}, \dfrac{2}{3}, \dfrac{1}{6}$ lcd _____

24. $\dfrac{31}{65}, \dfrac{6}{13}, \dfrac{2}{5}, \dfrac{7}{10}$ lcd _____

Show an estimate. Then show whether your estimate should be increased or decreased by circling the appropriate symbol. Do not adjust your estimate or compute the exact answer.

25. $\dfrac{3}{8} + \dfrac{6}{7} + \dfrac{11}{12} + \dfrac{5}{16}$

26. $\dfrac{2}{9} + \dfrac{3}{4} + \dfrac{7}{8} + \dfrac{7}{10}$

27. $5\dfrac{4}{9} + \dfrac{2}{3} + 6\dfrac{1}{2} + 2\dfrac{4}{5}$

28. $4\dfrac{1}{6} + 3\dfrac{3}{8} + \dfrac{5}{6} + 8\dfrac{1}{4}$

Add and show each sum in lowest terms.

29. $\dfrac{3}{15} + \dfrac{4}{5} + \dfrac{2}{3} =$

30. $11\dfrac{1}{9} + 12\dfrac{1}{2} + 14\dfrac{5}{18} =$

31. $\dfrac{5}{12} + \dfrac{3}{8} + \dfrac{3}{4} =$

32. $98\dfrac{7}{24} + 85\dfrac{3}{8} + 74\dfrac{17}{36} =$

17. _____
18. _____
19. _____
20. _____
21. _____
22. _____
23. _____
24. _____
25. _____ + −
26. _____ + −
27. _____ + −
28. _____ + −
29. _____
30. _____
31. _____
32. _____

3 Cumulative Review

Subtract and show the differences in lowest terms.

33. $\dfrac{8}{9} - \dfrac{3}{4} =$

34. $31\dfrac{1}{2} - 16\dfrac{1}{9} =$

35. $\dfrac{11}{15} - \dfrac{3}{8} =$

36. $92\dfrac{1}{15} - 62\dfrac{9}{10} =$

Multiply and show each product in lowest terms.

37. $\dfrac{9}{10} \times \dfrac{2}{3} \times \dfrac{3}{5} \times \dfrac{5}{7} =$

38. $21 \times 4\dfrac{2}{3} =$

39. $\dfrac{7}{8} \times \dfrac{2}{4} \times \dfrac{5}{6} \times \dfrac{12}{14} =$

40. $8\dfrac{9}{12} \times 9\dfrac{3}{7} =$

Divide and show each quotient in lowest terms.

41. $\dfrac{3}{8} \div \dfrac{2}{5} =$

42. $14 \div \dfrac{2}{7} =$

43. $12\dfrac{3}{8} \div 18 =$

44. $\dfrac{9}{11} \div 12 =$

45. $14\dfrac{4}{9} \div \dfrac{5}{6} =$

46. $54\dfrac{5}{12} \div 25\dfrac{5}{6} =$

Simplify these complex fractions to lowest terms.

47. $\dfrac{\frac{2}{3}}{17} =$

48. $\dfrac{\frac{3}{8}}{\frac{7}{12}} =$

49. $\dfrac{9\frac{6}{7}}{\frac{3}{5}} =$

33. _____
34. _____
35. _____
36. _____
37. _____
38. _____
39. _____
40. _____
41. _____
42. _____
43. _____
44. _____
45. _____
46. _____
47. _____
48. _____
49. _____

3 Cumulative Review

Solve these problems. Show the answers in lowest terms.

50. How much is $\frac{5}{6}$ of 48?

51. Four-fifths is what part of $\frac{5}{8}$?

52. How much is $\frac{7}{12}$ of 32?

53. Eighteen is what part of 40?

54. Fifty-two is $\frac{4}{9}$ of what number?

55. What part of 86 is 64.

56. Thirty-nine is three-fourths of what number?

57. At 45 bushels per acre, what is the yield of $1\frac{1}{4}$ acres?

58. If a horse eats $\frac{3}{8}$ bushel of oats each day, how many days will it take to eat 30 bushels of oats?

59. A diamond ring is valued at $3,250. If the diamond is valued at $7\frac{1}{3}$ times the value of the setting, what is the value of the setting?

60. Frank Gorsuch received $9,450 as full settlement of his claim against a bankrupt company. He received $\frac{5}{6}$ of his total claim. How much was his total claim?

61. Barbara Cerny bought $267\frac{1}{2}$ acres of farmland. Later she bought the adjoining farm consisting of $146\frac{5}{8}$ acres and then sold $187\frac{3}{4}$ acres from the two properties. How many acres of this land does she have left?

62. James Sumida, the executor, collected $315,350 and paid out expenses of $43,850 for an estate. In accordance with the will, he then distributed the balance to the widow and her four children as follows: $\frac{1}{3}$ to the widow and an equal portion of the remainder to each child. How much did each child receive?

63. A special blend of coffee was prepared that contained the following kinds and amounts: $14\frac{1}{2}$ pounds, Arabic; 29 pounds, Brazilian; and $21\frac{3}{4}$ pounds, Columbian. The Brazilian coffee made up what fractional part of the blend?

64. Ramon Morales, a furniture upholsterer, uses $3\frac{1}{3}$ yards of fabric to cover a certain chair. He has 5 bolts of fabric in various patterns. Each bolt contains 40 yards of fabric. How many such chairs can he cover with the fabric in stock?

65. Mrs. Debra Kong owned $\frac{3}{4}$ of a company and sold $\frac{2}{5}$ of her equity for $60,000. At this value, what is the value of the company?

LESSON 53

Reading and Rounding Decimals

A *decimal* is a fraction in which the denominator is understood to be a multiple of 10. A *mixed decimal* is a number composed of a whole number and a decimal. A dot, the *decimal point*, is placed to the right of the ones digit to show where the decimal starts. The first place to the right of the decimal point is valued at $\frac{1}{10}$ of 1; the second place $\frac{1}{10}$ of $\frac{1}{10}$ which is $\frac{1}{100}$ or 0.01. The figure to the right illustrates this. Notice that the names of the successive places to the right of the ones digit correspond to the names to the left.

To round decimal fractions, use the rules shown in Exercise 2 except that the digits to the right are dropped.

Example A

Thousands	Hundreds	Tens	Ones	DECIMAL POINT	Tenths	Hundredths	Thousandths
1	1	1	1	.	.1	.1	.1
($\frac{1}{10}$ of 10,000)	($\frac{1}{10}$ of 1,000)	($\frac{1}{10}$ of 100)	($\frac{1}{10}$ of 10)		($\frac{1}{10}$ of 1)	($\frac{1}{10}$ of $\frac{1}{10}$)	($\frac{1}{10}$ of $\frac{1}{100}$)

Example B Round 519.072364 to nearest ten thousandth.

519.072<u>3</u>64   Test digit (6) is 5 or larger.
     +1   Place digit (3) is increased by 1,
519.0724   and the six and four are dropped.

EXERCISE 53/Basic Time–3 Minutes

Estimated time to obtain a basic score of 100.

Answer these questions. (10 each)

1. Moving the decimal point two places to the left in a number has the effect of making the number (larger/smaller) ? .
2. Which numeral represents the larger value: 44 or 43.985?

Indicate whether the statement is true or false.

3. A decimal represents a number that is actually a fraction. This means that the denominator of the fraction is 10 or some power of 10.

Show the denominator for each of the following. Write your answers in the appropriate spaces.

4. 0.001 means $\frac{1}{?}$
5. 0.0001 means $\frac{1}{?}$
6. 0.000375 means $\frac{375}{?}$
7. 0.00097 means $\frac{97}{?}$

Use whole numbers and decimals to write the following.

8. 345 tens =
9. 452 tenths =
10. 38 thousands =
11. 47 thousandths =
12. Nine thousand forty-four ten-thousandths =

Round the following numerals as indicated.

13. 375.7986 to hundredths
14. 6,147.6728 to hundreds
15. 3,122.77026 to tens
16. 788.45435 to tenths
17. 107.106245 to thousandths
18. 390.152865 to ten-thousandths
19. $6.37125 to nearest cent
20. $0.096666 to nearest cent

1. smaller
2. 44
3. I guess True
4. 1000
5. 1/10,000
6. 10,000
7. 100,000
8. 3450
9. .452
10. 38,000
11. .047
12. 9000.0044
13. 375.8
14.
15. 3120
16. 788.5
17. 107.106
18. 390.1529
19. 6.30
20. 0.10

117

Name _____   Total Score _____
Date _____   Basic Score 100
Basic Time–12 Minutes   Improvement Score _____

Answer these questions. (8 each)

1. Moving the decimal point two places to the right in a number has the effect of making the number (larger/smaller) __?__.

2. Which numeral represents the larger value: 55 or 54.958?

For each of the following, indicate whether the statement is true or false.

3. A decimal represents a number that is actually a fraction. This means that the denominator of the fraction is 10 or some power of 10.

4. Appending zeros to the right of a decimal fraction changes the value of that decimal fraction.

Show the denominator for each of the following. Write your answers in the appropriate spaces.

5. 0.01 means $\frac{1}{?}$
6. 0.0001 means $\frac{1}{?}$
7. 0.0037 means $\frac{37}{?}$
8. 0.000278 means $\frac{278}{?}$

Use whole numbers and decimals to write the following.

9. 273 hundreds =
10. 342 hundredths =
11. 57 thousands =
12. 89 thousandths =
13. Seven thousand thirty-three ten-thousandths =
14. Five hundred eighty and sixty-seven hundred-thousandths =

Round the following numerals as indicated.

15. 357.8976 to hundredths
16. 5,238.5805 to hundreds.
17. 4,095.61087 to tenths
18. 7,198.54236 to tens
19. 216.305847 to thousandths
20. 480.215836 to ten-thousandths
21. 25.61374841 to millionths
22. 0.21583649 to ten-millionths
23. $3.62524 to nearest cent
24. $0.194675 to nearest cent
25. 5,930.813206584 to nearest ten-thousandth.

1. larger
2. _____
3. _____
4. _____
5. 1/100
6. 1/10,000
7. 37/10,000
8. 278/10,000,000
9. 273.00
10. .0342
11. 57,000
12. .089
13. 7,000.0033
14. _____
15. _____
16. _____
17. _____
18. _____
19. _____
20. _____
21. _____
22. _____
23. _____
24. _____
25. _____

118   © Copyright South-Western Publishing Co.

LESSON 54: Equivalent Decimal and Common Fractions

Study these examples that show how to change decimal fractions to equivalent common fractions and common fractions to decimal fractions.

Example A $0.0325 = \dfrac{325}{10,000} = \dfrac{13}{400}$

Solution A For the numerator of the result, write the decimal as a whole number. For the denominator, write 1, followed by as many 0's as there are places in the original decimal. Change to lowest terms.

Example B $\dfrac{1}{2} = 2\overline{)1.0}\;\;\dfrac{0.5}{} = 0.5$

Solution B Divide the numerator 1 by the denominator 2. To make this possible, put a decimal point and a 0 after the 1, giving 10 tenths; that is, 1.0. Dividing by 2 now gives 5 tenths or 0.5.

Example C $\dfrac{3}{125} = 125\overline{)3.000}\;\;\dfrac{0.024}{} = 0.024$

Solution C Put a decimal point and two 0's after the 3, making 300 hundredths; that is, 3.00. Now divide by 125, giving 2 hundredths or 0.02. As there is a remainder, supply additional 0's and continue the division.

EXERCISE 54/Basic Time–13 Minutes

Estimated time to obtain a basic score of 100.

Change to common fractions in lowest terms. Place each answer in the appropriate space on the right. (8 each)

1. $0.15 =$
2. $0.04 =$
3. $0.125 =$
4. $0.012 =$
5. $0.375 =$
6. $0.0003 =$
7. $0.01275 =$
8. $0.1272 =$
9. $0.0025 =$
10. $0.02075 =$
11. $0.6336 =$
12. $0.0378 =$

Change to equivalent decimal fractions. Round at the sixth decimal place where applicable. (8 each)

13. $\dfrac{3}{8} =$
14. $\dfrac{7}{16} =$
15. $\dfrac{20}{25} =$
16. $\dfrac{25}{32} =$
17. $\dfrac{5}{64} =$
18. $\dfrac{79}{125} =$
19. $\dfrac{2}{128} =$
20. $\dfrac{3}{250} =$
21. $\dfrac{16}{500} =$
22. $\dfrac{5}{7} =$
23. $\dfrac{13}{15} =$
24. $\dfrac{5}{124} =$
25. $\dfrac{3}{440} =$

1. $\dfrac{15}{100}$
2. $\dfrac{4}{100}$
3. $\dfrac{125}{1,000}$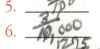
4. $\dfrac{12}{1,000}$
5. $\dfrac{375}{1,000}$
6. $\dfrac{3}{10,000}$
7. $\dfrac{1275}{100,000}$
8. $\dfrac{1272}{10,000}$
9. $\dfrac{25}{10,000}$
10. $\dfrac{2075}{100,000}$
11. $\dfrac{6336}{10,000}$
12. $\dfrac{378}{10,000}$
13. _____
14. _____
15. _____
16. _____
17. _____
18. _____
19. _____
20. _____
21. _____
22. _____
23. _____
24. _____
25. _____

TEST 54

Name _____ Total Score _____

Date _____ Basic Score __100__

Basic Time–9 Minutes Improvement Score _____

Change to common fractions in lowest terms. Place each answer in the appropriate space on the right. (8 each)

1. $0.25 =$

2. $0.05 =$

3. $0.025 =$

4. $0.005 =$

5. $0.125 =$

6. $0.0002 =$

7. $0.0125 =$

8. $0.625 =$

9. $0.00025 =$

10. $0.00125 =$

11. $0.3125 =$

12. $0.0875 =$

Change to equivalent decimal fractions. Round at the sixth decimal place where applicable. Place each answer in the appropriate space on the right. (8 each)

13. $\dfrac{5}{8} =$

14. $\dfrac{3}{16} =$

15. $\dfrac{9}{100} =$

16. $\dfrac{13}{25} =$

17. $\dfrac{15}{32} =$

18. $\dfrac{47}{125} =$

19. $\dfrac{13}{128} =$

20. $\dfrac{3}{1000} =$

21. $\dfrac{1}{50} =$

22. $\dfrac{2}{3} =$

23. $\dfrac{15}{16} =$

24. $\dfrac{7}{27} =$

25. $\dfrac{17}{45} =$

1. _____
2. _____
3. _____
4. _____
5. _____
6. _____
7. _____
8. _____
9. _____
10. _____
11. _____
12. _____
13. _____
14. _____
15. _____
16. _____
17. _____
18. _____
19. _____
20. _____
21. _____
22. _____
23. _____
24. _____
25. _____

LESSON 55: Decimal Equivalents

EXERCISE 55/Basic Time—7 Minutes

Estimated time to obtain a basic score of 100.

Supply decimal fractions for the fifty common fractions listed below. Show the decimal fractions as hundredths with remainders written as fractions. For example, $\frac{1}{3} = 0.33\frac{1}{3}$. (4 each)

NO.	COMMON FRACTION	DECIMAL FRACTION	NO.	COMMON FRACTION	DECIMAL FRACTION
1.	$\frac{1}{3}$		26.	$\frac{8}{9}$	
2.	$\frac{2}{3}$		27.	$\frac{1}{10}$	
3.	$\frac{1}{4}$		28.	$\frac{3}{10}$	
4.	$\frac{3}{4}$		29.	$\frac{7}{10}$	
5.	$\frac{1}{5}$		30.	$\frac{9}{10}$	
6.	$\frac{2}{5}$		31.	$\frac{1}{11}$	
7.	$\frac{3}{5}$		32.	$\frac{2}{11}$	
8.	$\frac{4}{5}$		33.	$\frac{1}{12}$	
9.	$\frac{1}{6}$		34.	$\frac{5}{12}$	
10.	$\frac{5}{6}$		35.	$\frac{7}{12}$	
11.	$\frac{1}{7}$		36.	$\frac{11}{12}$	
12.	$\frac{2}{7}$		37.	$\frac{1}{13}$	
13.	$\frac{3}{7}$		38.	$\frac{1}{14}$	
14.	$\frac{4}{7}$		39.	$\frac{1}{15}$	
15.	$\frac{5}{7}$		40.	$\frac{2}{15}$	
16.	$\frac{6}{7}$		41.	$\frac{1}{16}$	
17.	$\frac{1}{8}$		42.	$\frac{3}{16}$	
18.	$\frac{3}{8}$		43.	$\frac{5}{16}$	
19.	$\frac{5}{8}$		44.	$\frac{7}{16}$	
20.	$\frac{7}{8}$		45.	$\frac{11}{16}$	
21.	$\frac{1}{9}$		46.	$\frac{13}{16}$	
22.	$\frac{2}{9}$		47.	$\frac{15}{16}$	
23.	$\frac{4}{9}$		48.	$\frac{1}{20}$	
24.	$\frac{5}{9}$		49.	$\frac{1}{25}$	
25.	$\frac{7}{9}$		50.	$\frac{1}{40}$	

© Copyright South-Western Publishing Co.

TEST 55

Name _____ Total Score _____
Date _____ Basic Score __100__
Basic Time–8 Minutes Improvement Score _____

Fill in all spaces with the equivalents called for in the column headings. Show the common fractions in lowest terms and the decimal fractions as hundredths with remainders written as fractions. (4 each)

NO.	COMMON FRACTION	DECIMAL FRACTION	NO.	COMMON FRACTION	DECIMAL FRACTION
1.		1.50	26.		$0.12\frac{1}{2}$
2.		$0.02\frac{1}{2}$	27.	$\frac{9}{20}$	
3.	$\frac{3}{50}$		28.	$\frac{4}{7}$	
4.	$1\frac{1}{4}$		29.		$0.14\frac{2}{7}$
5.		$0.07\frac{1}{2}$	30.	$\frac{1}{200}$	
6.	$\frac{5}{20}$		31.		$0.83\frac{1}{3}$
7.		0.05	32.		0.80
8.		$0.93\frac{3}{4}$	33.	$\frac{3}{20}$	
9.	$\frac{1}{100}$		34.	$\frac{6}{10}$	
10.	$\frac{5}{80}$		35.		1.20
11.		$0.16\frac{2}{3}$	36.	$1\frac{1}{16}$	
12.	$2\frac{1}{5}$		37.		0.73
13.		$0.08\frac{1}{3}$	38.		0.25
14.		$0.09\frac{1}{11}$	39.	$1\frac{5}{8}$	
15.	$\frac{7}{20}$		40.	$\frac{97}{100}$	
16.	$\frac{9}{10}$		41.		$0.66\frac{2}{3}$
17.		0.70	42.	$1\frac{3}{10}$	
18.	$\frac{4}{25}$		43.		1.90
19.		$0.88\frac{8}{9}$	44.		3.50
20.		$0.11\frac{1}{9}$	45.	$2\frac{3}{5}$	
21.	$\frac{1}{800}$		46.	$10\frac{1}{2}$	
22.	$\frac{7}{8}$		47.		$0.33\frac{1}{3}$
23.		$0.62\frac{1}{2}$	48.	$1\frac{3}{4}$	
24.	$\frac{1}{50}$		49.		1.05
25.		$0.37\frac{1}{2}$	50.	$\frac{5}{6}$	

© Copyright South-Western Publishing Co.

LESSON 56: Estimation with Decimals

Decimal notation is simply another way to write fractions. As with Exercise 40, completing this exercise will improve your understanding of fractional size. This will help you decide whether your answers to problems containing decimals "make sense."

EXERCISE 56/Basic Time—4 Minutes
Estimated time to obtain a basic score of 100.

In 1-12, each problem offers three choices. Choose only one. Write your answers in the appropriate spaces on the right. (10 each)

1. 0.47021 is near 0, $\frac{1}{2}$, or 1.
2. 0.93 is near 0, $\frac{1}{2}$, or 1.
3. 0.035402 is near 0, $\frac{1}{2}$, or 1.
4. 0.87815 is near 0, $\frac{1}{2}$, or 1.
5. 0.48314 is near 0, $\frac{1}{2}$, or 1.
6. 0.79268 is near $\frac{1}{4}$, $\frac{1}{2}$, or $\frac{3}{4}$.
7. 0.09321 is near $\frac{1}{100}$, $\frac{1}{10}$, or 1.
8. 0.8456 is near $\frac{1}{100}$, $\frac{1}{10}$, or 1.
9. 0.00794 is near $\frac{1}{100}$, $\frac{1}{10}$, or 1.
10. 0.24356 is near $\frac{1}{4}$, $\frac{1}{2}$, or $\frac{3}{4}$.
11. 0.0819 is near 1, $\frac{1}{10}$, or $\frac{1}{100}$.
12. 0.00873 is near 1, $\frac{1}{10}$, or $\frac{1}{100}$.

Don't compute the exact answer. In the appropriate space on the right, write an estimate for each problem. (10 each)

13. $2.16
    1.47
    0.98
    + 4.65

14. $10.28
    5.62
    2.30
    + 1.49

15. 6.84
    3.705
    0.45
    + 10.52

16. .035
    × 0.093

17. 247
    × 0.487

18. 6.295
    × 0.2356

19. 8.00000
    − 0.73014

20. 0.0976) 25

1. _____
2. _____
3. _____
4. _____
5. _____
6. _____
7. _____
8. _____
9. _____
10. _____
11. _____
12. _____
13. _____
14. _____
15. _____
16. _____
17. _____
18. _____
19. _____
20. _____

TEST 56

Name _____ Total Score _____
Date _____ Basic Score __100__
Basic Time–4 Minutes Improvement Score _____

In 1-12, each problem offers three choices. Choose only one. Write your answers in the appropriate spaces on the right. (10 each)

1. 0.02521 is near 0, $\frac{1}{2}$, or 1.

2. 0.94 is near 0, $\frac{1}{2}$, or 1.

3. 0.485402 is near 0, $\frac{1}{2}$, or 1.

4. 0.47815 is near 0, $\frac{1}{2}$, or 1.

5. 0.88314 is near 0, $\frac{1}{2}$, or 1.

6. 0.26268 is near $\frac{1}{4}$, $\frac{1}{2}$, or $\frac{3}{4}$.

7. 0.89321 is near $\frac{1}{100}$, $\frac{1}{10}$, or 1.

8. 0.0956 is near $\frac{1}{100}$, $\frac{1}{10}$, or 1.

9. 0.00974 is near $\frac{1}{100}$, $\frac{1}{10}$, or 1.

10. 0.23569 is near $\frac{1}{4}$, $\frac{1}{2}$, or $\frac{3}{4}$.

11. 0.00819 is near 1, $\frac{1}{10}$, or $\frac{1}{100}$.

12. 0.0873 is near 1, $\frac{1}{10}$, or $\frac{1}{100}$.

1. _____
2. _____
3. _____
4. _____
5. _____
6. _____
7. _____
8. _____
9. _____
10. _____
11. _____
12. _____

Don't compute the exact answer. In the appropriate space on the right, write an estimate for each problem. (10 each)

13. $ 6.12
    7.14
    0.89
  + 5.46

14. $ 15.82
    2.56
    3.20
  + 9.14

15.   4.86
    5.30
    0.757
  + 20.26

16.   54
    × 0.323

17.   724
    × 0.085

18.   5926
    × 0.4756

19.   8
   − 0.06437

20. 0.8092)̄65

13. _____
14. _____
15. _____
16. _____
17. _____
18. _____
19. _____
20. _____

124 © Copyright South-Western Publishing Co.

LESSON 57

Placement of the Decimal Point

Since the proper placement of the decimal point seems to cause trouble, its placement in addition, subtraction, and multiplication is considered in this exercise. Division of decimals is discussed in Exercise 59.

When adding and subtracting decimals, align the decimal points, thus keeping each digit in its own column. See the addition problem in Example A. Attach zeros in the minuend for subtraction, if desired.

When multiplying decimals, multiply them as whole numbers. Then mark off in the product the same number of decimal places as found in both the multiplier and multiplicand. See Example B.

Example A Add 3.14 + 18.4 + 340.1:

```
   3.14
  18.40
 340.10
 ──────
 361.64
```

Example B Multiply 40.8 × 3.02:

```
  40.8      1 decimal place
 x3.02      2 decimal places
 ─────
   816
 122 40
 ──────
 123.216    3 decimal places
```

EXERCISE 57/Basic Time – 10 Minutes

Estimated time to obtain a basic score of 100.

Add:

1.  ```
    325.025
     36.05
      0.3684
      1.5
    225.0725
      6.1728
     43.375
    ```
    (10)

2. 3.6024 + 18.32 + 51.05 + 16.5 + 187.16 + 0.0275 = _____ (24)

3. 187.1 + 0.09275 + 3.875 + 0.005 + 18.75 + 62.0045 = _____ (24)

4. 0.02768 + 5.125 + 1,376.0048 + 62.476 + 3.2 + 5.84 = _____ (24)

5. 90 + 0.028 + 3.005 + 76 + 0.02 + 15.8 + 1.1652 = _____ (24)

Subtract:

6.  ```
    27.3476
     8.09
    ```
    (4)

7.  ```
    14.02500
     0.97875
    ```
    (4)

8. 38.1 − 4.56789 = _____ (6)
9. 257 − 152.0672 = _____ (6)
10. 0.0947 − 0.00965 = _____ (6)
11. 13.4786 − 2.75 = _____ (6)
12. 4.92 − 3.8769 = _____ (6)
13. 125 − 27.00912 = _____ (6)
14. 0.57 − 0.001027 = _____ (6)

Multiply:

15. ```
    26.535
    385.02
    ```
    (16)

16. ```
    15.376
     4.27
    ```
    (14)

17. ```
    14.267
     0.0095
    ```
    (14)

TEST 57

Name _____ Total Score _____
Date _____ Basic Score __100__
Basic Time–10 Minutes Improvement Score _____

Add:

1. 324.3025 + 0.02 + 34.786 + 1.00055 + 0.375 + 6.248 + 16.01 + 0.025 + 14.25 = _____ (28)

2. 17.2 + 185 + 0.0384 + 8 + 7.478 + 0.025 = _____ (20)

3. 0.0274 + 3.8 + 165 + 24.5675 + 0.0027 = _____ (20)

4. 6.2784 + 0.02 + 3.875 + 14.1 + 16.01 + 5.5 = _____ (20)

5. 2.005 + 0.0375 + 15.1875 + 7.85 + 0.025 = _____ (20)

Subtract: (4 each)

6. 13.025 − 0.9875 = _____

7. 6.76 − 4.9725 = _____

8. 54.1 − 4.8257 = _____

9. 7.548 − 0.9786 = _____

10. 725 − 0.0937 = _____

11. 80.24 − 0.9876 = _____

12. 4.21 − 2.987 = _____

13. 6,421 − 0.0247 = _____

Multiply: (10 each)

14. 0.00487
     0.00421

15. 5,874
     0.008

16. 56.021
     0.0078

17. 0.6529
     0.426

18. 0.000371
     0.424

19. 72.412
     725

126 © Copyright South-Western Publishing Co.

LESSON 58

Figuring Gross Pay

One practical use of your knowledge of multiplying with decimals is in figuring your earnings. In business, such knowledge is used in the preparation of a payroll. The gross pay portion of a payroll is shown below.

In this exercise, add each horizontal row of figures under HOURS WORKED DAILY and place the sum on the line under TOTAL HOURS WEEK. Next multiply each wage per hour by the number of hours worked by the employee. Enter the result in the column headed GROSS WAGES. Then total that column.

EXERCISE 58/Basic Time–8 Minutes

Estimated time to obtain a basic score of 100.

Complete the following payroll: (4 each, Total Hrs. Week column; 12 each, Gross Wages column; 8, Total)

PAYROLL   Date _____

EMPLOYEE	HRS. WORKED DAILY						TOTAL HRS. WEEK	WAGES PER HR.		GROSS WAGES	
	M	T	W	T	F	S					
Chrisman, W.	8	8	8	8	8	0		8	55		
Donosky, J.	8	8	6	8	8	0		15	90		
Fernandez, P.	8	8	8	8	8	0		9	10		
Glover, J.	0	8	8	8	8	0		8	25		
Helm, G. G.	8	8	7	8	8	0		16	50		
Hillman, E. P.	8	7	6	8	8	0		12	20		
Juarez, E.	8	8	8	8	8	0		9	97		
Klein, A. A.	8	7	5	8	8	0		8	95		
Liang, C.	8	8	5	8	8	0		14	25		
Peek, K.	8	8	8	8	7	0		17	50		
Pritchard, M.	8	8	4	0	8	0		10	40		
Towers, H.	8	8	4	8	8	0		8	30		
									Total		

TEST 58

Name _____ Total Score _____

Date _____ Basic Score _100_

Basic Time–16 Minutes Improvement Score _____

Complete the following payroll: (2 each, Total Hrs. Week column; 6 each, Gross Wages column; 8, Total)

	PAYROLL							Date _____	
EMPLOYEE	HRS. WORKED DAILY						TOTAL HRS. WEEK	WAGES PER HR.	GROSS WAGES
	M	T	W	T	F	S			
Aiken, D.	8	8	8	8	8	0		8 10	
Bernstein, L.	8	8	8	4	8	0		15 10	
Campbell, J.	0	4	8	8	8	0		12 25	
Chancey, W.	8	8	8	8	8	0		9 05	
Crisp, M.	8	8	8	8	8	0		16 50	
Griffiths, M.	8	8	8	5	8	0		9 97	
Guerra, M.	8	8	8	8	7	0		14 97	
Horn, A. B.	8	4	8	8	8	0		8 85	
Ing, C.	4	8	8	8	7	0		16 95	
Jensen, J.	8	8	8	6	8	0		9 05	
Klem, M. S.	8	8	7	7	8	0		10 40	
Laredo, C.	8	8	8	8	5	0		9 30	
Malkovich, D.	5	8	8	8	8	0		12 20	
Newsome, C.	8	8	6	8	8	0		10 25	
Orson, W. A.	8	8	8	4	8	0		17 15	
Ponsetti, J.	8	8	8	8	8	0		12 75	
Powell, E.	8	8	8	8	8	0		8 52	
Ryan, C.	8	2	8	8	8	0		15 10	
Steinberg, B.	8	8	8	8	1	0		12 75	
Thomas, R.	8	8	8	8	7	0		9 99	
Ultman, G.	8	7	7	7	7	0		11 99	
Vergari, D.	8	8	8	8	4	0		9 15	
Walken, G.	8	8	8	8	8	0		16 85	
Young, V. C.	8	8	3	8	8	0		14 20	
								Total	

© Copyright South-Western Publishing Co.

LESSON 59
Division of Decimals

To divide decimals, divide as in whole numbers. But before starting the division, move the decimal point in the divisor enough places to the right to make it a whole number. Next move the decimal point in the dividend the same number of places, attaching 0's if necessary. Place the decimal point for the quotient directly above the new position of the decimal point in the dividend. Then proceed to divide. The division may be continued by attaching as many 0's as desired.

```
                              3.7    quotient
         divisor   2.25.)8.32.5     dividend
                              6 75
                              1 575
                              1 575
```

EXERCISE 59/Basic Time—7 Minutes
Estimated time to obtain a basic score of 100.

Divide. Round at the fifth decimal place where applicable. Place each answer in the appropriate space on the right. (20 each, Problems 1-4; 30 each, Problems 5-8)

1. 12.71)185.566   2. 85.14)147.287   1. _____

2. _____

3. 26.4)1,071.84   4. 42.5)0.00978   3. _____

4. _____

5. _____

5. 78.75)0.025   6. 3,275)1.15   6. _____

7. _____

7. 0.567)68.4   8. 0.0547)78.   8. _____

Name	Total Score
Date	Basic Score 100
Basic Time–5 Minutes	Improvement Score

Divide. Round at the fifth decimal place where applicable. Place each answer in the appropriate space on the right. (20 each)

1. 2.5)0.0635   2. 0.00025)38   1._____

2._____

3. 87.6)999.006   4. 387)0.0062   3._____

4._____

5. 4.36)0.001   6. 0.000125)50   5._____

6._____

7. 0.0625)0.12   8. 0.364)109.2   7._____

8._____

9. 0.015)1   10. 0.07)1.3265   9._____

10._____

130 © Copyright South-Western Publishing Co.

LESSON 60: Shortcut Division by 10 and by Multiples of 10

To divide by 10, 100, 1,000, etc., move the decimal point in the dividend to the left as many places as there are 0's in the divisor. When necessary, attach 0's on the left as place holders. See Example A.

Example A

365 ÷	10 =	36.5	point moved 1 place to left
365 ÷	100 =	3.65	point moved 2 places to left
365 ÷	1,000 =	0.365	point moved 3 places to left
365 ÷	10,000 =	0.0365	point moved 4 places to left

This rule may be extended to divide rapidly by any number ending in one or more 0's, such as 40, 700, and 3,000. To divide by any number ending in one or more 0's: (1) move the decimal point in the divisor to the left the number of places necessary to eliminate its 0's; (2) move the decimal point in the dividend the same number of places to the left; (3) then proceed to divide. See Example B.

Example B

```
         0.037              0.037
700)25.90      or     700) .25.9
     21 00                   21
      4 900                   49
      4 900                   49
```

When both the divisor and the dividend are divided by the same number (except 0), the value of the quotient remains unchanged. Thus, when both the divisor and the dividend in Example B are divided by 100, the size of the divisor and dividend is decreased, but the value of the quotient remains unchanged. Dividing by the decreased divisor is easier.

EXERCISE 60/Basic Time—8 Minutes
Estimated time to obtain a basic score of 100.

Divide. Round at the fourth decimal place where applicable. (8 each)

1. 345 ÷ 10 = _____
2. 376 ÷ 20 = _____
3. 393 ÷ 30 = _____
4. 367 ÷ 100 = _____
5. 303 ÷ 200 = _____
6. 382.2 ÷ 300 = _____
7. 350 ÷ 1,000 = _____
8. 483 ÷ 2,000 = _____
9. 40.5 ÷ 3,000 = _____
10. 4.16 ÷ 60 = _____
11. 492 ÷ 400 = _____
12. 54.4 ÷ 8,000 = _____
13. 0.653 ÷ 500 = _____
14. 56.4 ÷ 70 = _____
15. 6,055 ÷ 600 = _____
16. 607 ÷ 400 = _____
17. 61.3 ÷ 5,000 = _____
18. 7.829 ÷ 8,000 = _____
19. 740 ÷ 900 = _____
20. 0.873 ÷ 50 = _____
21. 80.8 ÷ 90 = _____
22. 9.254 ÷ 700 = _____
23. 90.96 ÷ 60.00 = _____
24. 91.30 ÷ 40.00 = _____
25. 119.7 ÷ 70.00 = _____

© Copyright South-Western Publishing Co.

TEST 60

Name _____ Total Score _____
Date _____ Basic Score __100__
Basic Time–15 Minutes Improvement Score _____

Divide. Round at the fourth decimal place where applicable. (4 each)

1. 34 ÷ 10 = _____
2. 96 ÷ 100 = _____
3. 98 ÷ 1,000 = _____
4. 31 ÷ 20 = _____
5. 35.7 ÷ 900 = _____
6. 64 ÷ 200 = _____
7. 48 ÷ 3,000 = _____
8. 47 ÷ 10 = _____
9. 84 ÷ 70 = _____
10. 57 ÷ 100.00 = _____
11. 57 ÷ 40 = _____
12. 65 ÷ 100 = _____
13. 63 ÷ 3,000 = _____
14. 66.5 ÷ 70 = _____
15. 57 ÷ 300 = _____
16. 79 ÷ 4,000 = _____
17. 85.4 ÷ 80 = _____
18. 89 ÷ 5,000 = _____
19. 9.36 ÷ 400 = _____
20. 10.17 ÷ 90 = _____
21. 1.07 ÷ 90 = _____
22. 211 ÷ 60.00 = _____
23. 110 ÷ 800.00 = _____
24. 373 ÷ 60 = _____
25. 80 ÷ 400 = _____
26. 357 ÷ 4,000 = _____
27. 67.3 ÷ 20 = _____
28. 393 ÷ 600 = _____
29. 43.92 ÷ 30 = _____
30. 2.49 ÷ 80 = _____
31. 43.1 ÷ 1,000 = _____
32. 57.47 ÷ 7,000 = _____
33. 510 ÷ 50.00 = _____
34. 354 ÷ 30 = _____
35. 632 ÷ 800 = _____
36. 465 ÷ 500.0 = _____
37. 681 ÷ 700 = _____
38. 728 ÷ 50 = _____
39. 478 ÷ 40.00 = _____
40. 862 ÷ 20,000 = _____
41. 823.8 ÷ 6,000 = _____
42. 48.72 ÷ 800 = _____
43. 9.906 ÷ 60 = _____
44. 105.8 ÷ 500 = _____
45. 208.16 ÷ 700.00 = _____
46. 1.159 ÷ 90 = _____
47. 0.768 ÷ 80.00 = _____
48. 3.564 ÷ 600.0 = _____
49. 7,368 ÷ 4,000 = _____
50. 39.097 ÷ 9,000 = _____

© Copyright South-Western Publishing Co.

LESSON 61 — Aliquot Parts

A number (whole or mixed) that divides into another number leaving no remainder is an aliquot part of that number. Therefore, 5 is an aliquot part of 10, 4 is an aliquot part of 12, $12\frac{1}{2}$ is an aliquot part of 100, and 25 is an aliquot part of 100. Looking at this relationship in another way, 5 is to 10 as 1 is to 2:

$$\frac{5}{10} = \frac{1}{2}$$

The relationships of the other numbers mentioned above are:

$$\frac{4}{12} = \frac{1}{3} \qquad \frac{12\frac{1}{2}}{100} = \frac{25}{200} = \frac{1}{8} \qquad \frac{25}{100} = \frac{1}{4}$$

Such relationships between numbers may be used to simplify the solution of a problem. An advantage of aliquot-part shortcut methods is that many problems may be solved mentally. The mental solution to the example below may be deduced in this manner: 56 pounds at $1 a pound would cost $56; but if the price were only $\frac{1}{8}$ of $1 for each pound, the cost would then be only $\frac{1}{8}$ as much, or $7 ($\frac{1}{8}$ of $56).

Example: 56 lb. @ $12\frac{1}{2}$¢ = ?

Solution: $56 \times \$0.12\frac{1}{2} = 56 \times \$\frac{1}{8} = \$7$

Notice that when the cent sign (¢) is removed, the decimal point is moved two places to the left.

EXERCISE 61 / Basic Time – 8 Minutes
Estimated time to obtain a basic score of 100.

Each of these aliquot parts has a base of 100. Show the common fraction equivalent for each. (10 each)

NO.	PROBLEM	SOLUTION	NO.	PROBLEM	SOLUTION
0.	$12\frac{1}{2}$ = $\frac{1}{8}$ of 100	$\frac{12\frac{1}{2}}{100} = \frac{25}{200} = \frac{1}{8}$	9.	40 = ___ of 100	
1.	10 = ___ of 100		10.	$87\frac{1}{2}$ = ___ of 100	
2.	20 = ___ of 100		11.	$37\frac{1}{2}$ = ___ of 100	
3.	25 = ___ of 100		12.	$14\frac{2}{7}$ = ___ of 100	
4.	75 = ___ of 100		13.	$18\frac{3}{4}$ = ___ of 100	
5.	$16\frac{2}{3}$ = ___ of 100		14.	$62\frac{1}{2}$ = ___ of 100	
6.	$33\frac{1}{3}$ = ___ of 100		15.	$83\frac{1}{3}$ = ___ of 100	
7.	$8\frac{1}{3}$ = ___ of 100		16.	$41\frac{2}{3}$ = ___ of 100	
8.	$6\frac{1}{4}$ = ___ of 100		17.	$11\frac{1}{9}$ = ___ of 100	

In the parentheses show the fractional part of $1 for each price and determine the total cost: (2 for each fraction; 4 for each total cost)

0. 64 @ 10¢ ($\frac{1}{10}$) = $ 6.40

18. 60 @ 5¢ (___) = $_____

19. 84 @ 25¢ (___) = $_____

20. 470 @ 50¢ (___) = $_____

21. 575 @ 20¢ (___) = $_____

22. 864 @ $16\frac{2}{3}$¢ (___) = $_____

TEST 61

Name _____ Total Score _____
Date _____ Basic Score __100__
Basic Time–10 Minutes   Improvement Score _____

Each of these aliquot parts has a base of 100. Show the common fraction equivalent. (4 each)

NO.	PROBLEM	SOLUTION	NO.	PROBLEM	SOLUTION
1.	25 = of 100		2.	$33\frac{1}{3}$ = of 100	
3.	60 = of 100		4.	$16\frac{2}{3}$ = of 100	
5.	$14\frac{2}{7}$ = of 100		6.	$12\frac{1}{2}$ = of 100	
7.	$6\frac{1}{4}$ = of 100		8.	$37\frac{1}{2}$ = of 100	
9.	$18\frac{3}{4}$ = of 100		10.	$62\frac{1}{2}$ = of 100	
11.	$58\frac{1}{3}$ = of 100		12.	$83\frac{1}{3}$ = of 100	
13.	$93\frac{3}{4}$ = of 100		14.	$66\frac{2}{3}$ = of 100	
15.	$87\frac{1}{2}$ = of 100		16.	75 = of 100	
17.	$8\frac{1}{3}$ = of 100		18.	$11\frac{1}{9}$ = of 100	
19.	$41\frac{2}{3}$ = of 100		20.	$91\frac{2}{3}$ = of 100	
21.	$9\frac{1}{11}$ = of 100		22.	$56\frac{1}{4}$ = of 100	

Each of these aliquot parts has a base of $1. Show the common fraction equivalent and determine the total cost. (4 for each fraction; 4 for each total cost)

23. 60 @ 50¢ ( ) = $_____   24. 84 @ 25¢ ( ) = $_____

25. 64 @ $12\frac{1}{2}$¢ ( ) = $_____   26. 84 @ $16\frac{2}{3}$¢ ( ) = $_____

27. 381 @ $33\frac{1}{3}$¢ ( ) = $_____   28. 375 @ 20¢ ( ) = $_____

29. 306 @ $16\frac{2}{3}$¢ ( ) = $_____   30. 91 @ $14\frac{2}{7}$¢ ( ) = $_____

31. 72 @ $11\frac{1}{9}$¢ ( ) = $_____   32. 588 @ $8\frac{1}{3}$¢ ( ) = $_____

33. 320 @ $6\frac{1}{4}$¢ ( ) = $_____   34. 480 @ 75¢ ( ) = $_____

35. 480 @ 60¢ ( ) = $_____   36. 705 @ 80¢ ( ) = $_____

© Copyright South-Western Publishing Co.

4 Cumulative Review

Answer these questions.

1. Moving the decimal point two places to the left in a number has the effect of making the number (larger/smaller) _?_ .

2. Moving the decimal point two places to the right in a number has the effect of making the number (larger/smaller) _?_ .

3. Which numeral represents the larger value: 66 or 65.985?

For each of the following, indicate whether the statement is *true* or *false*.

4. A decimal represents a number that is actually a fraction. This means that the denominator of the fraction is 10 or some power of 10.

5. Appending zeros to the right of a decimal fraction changes the value of that decimal fraction.

6. The addition of decimal fractions is exactly the same as the addition of whole numbers.

7. The multiplication of decimal fractions is the same as that of whole numbers except that the decimal point must be placed properly in the product.

8. When a divisor is a decimal, it may be changed to a whole number by multiplying it by some power of 10 if the dividend is also multiplied by the same power of 10.

9. When only the divisor is a decimal, the division is the same as that of whole numbers except that the decimal point must be placed in the quotient exactly above that in the dividend.

10. When only the dividend is a decimal, the division is the same as that of whole numbers after the decimal point in the dividend is moved the appropriate number of places to the right.

11. A common fraction may be changed to a decimal by dividing the denominator by the numerator.

12. To find the fractional equivalent of any aliquot part, place the aliquot part over its base and reduce the fraction to lowest terms.

Round the following numerals as indicated.

13. 357.8976 to hundredths

14. 5,238.5805 to hundreds

15. 0.094\frac{1}{2}$ to nearest cent

Change each of the following to a common fraction.

16. 0.12$\frac{1}{2}$

17. 0.58$\frac{1}{3}$

18. 0.93$\frac{3}{4}$

19. 23.37$\frac{1}{2}$

20. 9.06$\frac{1}{4}$

21. 45.33$\frac{1}{3}$

© Copyright South-Western Publishing Co.

4 Cumulative Review

Change each of the following to an equivalent decimal fraction in hundredths with remainders written as fractions. Do not round.

22. $\frac{1}{4}$ 23. $\frac{5}{8}$ 24. $\frac{1}{6}$

Change each of the following to an equivalent mixed decimal. Where applicable, round to the nearest ten-thousandth.

25. $4\frac{2}{3}$ 26. $275\frac{7}{8}$ 27. $90\frac{3}{5}$

Each problem offers three choices. Choose only one.

28. 0.87021 is near 0, $\frac{1}{2}$, or 1.

29. 0.09321 is near $\frac{1}{100}$, $\frac{1}{10}$, or 1.

30. 0.24356 is near $\frac{1}{4}$, $\frac{1}{2}$, or $\frac{3}{4}$.

Show an estimate for each problem. Do not compute the exact answer.

31. $3.25
     0.57
     5.23
    +1.72

32.   326
     × 0.487

33.   468
     × 0.3507

34.   37
     − 0.82041

35. $0.7342 \overline{)36}$

Solve these problems.

36. 24.32 + 7.862 + 0.0085 + 8.6 + 4.019 =

37. $5.48 + 76\frac{2}{5} + 38.32\frac{7}{10} + 8.29\frac{1}{4} =$

38. $3.50 − 89¢ =

39. $243.085 − 98\frac{5}{8} =$

40. 52.38 × 0.0096 =

41. $0.45\frac{7}{8} \times 74.12 =$

42. 459.7425 ÷ 48.65 =

43. To the nearest ten-thousandth, find the quotient of $82\frac{1}{2} \div 6.87\frac{1}{4} =$

22. _____
23. _____
24. _____
25. _____
26. _____
27. _____
28. _____
29. _____
30. _____
31. _____
32. _____
33. _____
34. _____
35. _____
36. _____
37. _____
38. _____
39. _____
40. _____
41. _____
42. _____
43. _____

4 Cumulative Review

Name _____

Complete the following payroll:

Employee	S	M	T	W	T	F	S	Total Hr/Week	Wages Per Hr	
44. Champeau, J.	0	8	8	8	8	8	0	_____	$8.90	44._____
45. Gordan, D.	8	8	0	7	8	7	0	_____	14.10	45._____
46. Moore, G.	8	0	0	8	5	8	8	_____	16.50	46._____
47. Okano, B.	0	0	7	8	8	5	8	_____	10.36	47._____
48. Perez, R.	0	7	7	0	7	6	8	_____	9.25	48._____
49. Ruemmler, W.	8	8	8	8	7	0	0	_____	15.78	49._____
50. Weinstein, C.	0	8	8	7	8	6	0	_____	12.96	50._____

To the nearest cent, find the total amount for each of the following:

51. 19,250 lb @ $84 per 2,000 lb =   51._____

52. 7,586 ft @ $35 per 1,000 ft =   52._____

53. 875 ft @ $27.95 per 100 ft =   53._____

54. 2,860 lb @ $61.50 per 100 lb =   54._____

Show the fractional equivalent for each of the following:

55. $\underline{\ ?\ } \times 100 = 8\frac{1}{3}$ 55._____

56. $\underline{\ ?\ } \times 100 = 83\frac{1}{3}$ 56._____

57. $\underline{\ ?\ } \times \$1 = 6\frac{1}{4}$ ¢   57._____

In cents, show the aliquot part of $1 that is represented by the following:

58. $\frac{1}{7}$ of $1 =   58._____

59. $\frac{1}{9}$ of $1 =   59._____

60. $\frac{1}{8}$ of $1 =   60._____

Use the aliquot-parts method to find the cost of each set.

61. 84 @ 75¢ each =   61._____

62. 132 @ $25 each =   62._____

63. 50 @ $34 each =   63._____

4 Cumulative Review

Solve these problems.

64. The manager of a store bought 22 boxes of cleaner at $37.50 a box. How much was the total cost?

64._____

65. A purchase of $8\frac{1}{2}$ yards of plastic material was made at $1.12 a yard. How much was the total cost?

65._____

66. How much will 25 pens cost at $7.98 each?

66._____

67. If 4 small pencils cost 75¢, how many pencils can be purchased for $5.25?

67._____

68. A certain steel beam weighs 18.4 pounds per foot of length. If this beam is 26.5 feet long, how many pounds does it weigh?

68._____

69. The total cost, including carrying charges, for the boat that Mr. Thomas purchased was $18,943.76. He paid $2,750 down and agreed to pay the balance in 12 equal monthly payments. How much was each payment?

69._____

70. Last year a family paid $10,500 for doctors' bills. If this was equal to 0.30 of the family's annual income, how much was the annual income?

70._____

71. Ruben Diaz set a goal to earn $500 commissions for the week before Christmas. Commissions for each sales day of the week were: $67.14, $76.29, $80.34, $76.10, $87.30, and $89.07. How much less than his goal did he earn in commissions?

71._____

72. An airplane was flown 1,298.3 miles in 3.75 hours. Find the average speed to the nearest tenth of a mile per hour.

72._____

73. A store bought 250 toys at $16.97 each to sell during the holidays. What was the total cost to the store?

73._____

74. Agnes McCray received commissions totaling $158.85 for having made 29 sales. To the nearest cent, what is the average amount of commission on each sale?

74._____

75. John Mako earns $8.75 an hour working after class and on Saturdays. He works $4\frac{1}{2}$ hours on each of five days and 8 hours on Saturday. How much is his gross weekly wage?

75._____

LESSON 62

Changing Percents to Decimals and Fractions

Percent means *hundredths*. Percents offer a way to write decimal fractions in which the denominator is understood to be 100. Five percent means five one-hundredths and may be written with numerals in three ways: $\frac{5}{100}$; 0.05, and 5%. *All have the same mathematical value.*

You should know how to change a percent to either a decimal or a common fraction.

1. To change a percent to a decimal fraction, drop the percent sign (%) and move the decimal point two places to the left.
2. To change a percent to a common fraction, first change the percent to a decimal and then to a common fraction in lowest terms.

Example: Change (a) 25% and (b) $33\frac{1}{3}$% to decimals.
(a) 25% = 0.25
(b) $33\frac{1}{3}$% = $0.33\frac{1}{3}$

Example: Change (a) 25% and (b) $12\frac{1}{2}$% to common fractions.
(a) 25% = 0.25 = $\frac{25}{100}$ = $\frac{1}{4}$
(b) $12\frac{1}{2}$% = $0.12\frac{1}{2}$ = 0.125 = $\frac{125}{1000}$ = $\frac{1}{8}$

EXERCISE 62/Basic Time–10 Minutes
Estimated time to obtain a basic score of 100.

Change each of the following percents to a decimal equivalent. (4 each)

	Decimal			Decimal
1. 40% =	1. _____	2. 5% =	2. _____
3. 135% =	3. _____	4. 2% =	4. _____
5. 275% =	5. _____	6. 80% =	6. _____
7. 0.4% =	7. _____	8. 225% =	8. _____
9. $\frac{1}{4}$% =	9. _____	10. 800% =	10. _____
11. $37\frac{1}{2}$% =	11. _____	12. 0.09% =	12. _____
13. $7\frac{1}{2}$% =	13. _____	14. 75% =	14. _____
15. 6.5% =	15. _____	16. $33\frac{1}{3}$% =	16. _____
17. 0.75% =	17. _____	18. $\frac{1}{8}$% =	18. _____

Change each of the following to decimal and common fraction equivalents. (4 each answer)

	Decimal Fraction	Common Fraction		Decimal Fraction	Common Fraction
19. 10% =	19. _____	_____	20. 40% =	20. _____	_____
21. 62.5% =	21. _____	_____	22. 875% =	22. _____	_____
23. 7.5% =	23. _____	_____	24. $37\frac{1}{2}$% =	24. _____	_____
25. 125% =	25. _____	_____	26. $\frac{1}{2}$% =	26. _____	_____
27. 75% =	27. _____	_____	28. 5% =	28. _____	_____
29. $66\frac{2}{3}$% =	29. _____	_____	30. 25% =	30. _____	_____
31. 0.9% =	31. _____	_____	32. $6\frac{1}{4}$% =	32. _____	_____
33. $8\frac{1}{3}$% =	33. _____	_____	34. $\frac{3}{4}$% =	34. _____	_____

TEST 62

Name _____ Total Score _____
Date _____ Basic Score 100
Basic Time–9 Minutes   Improvement Score _____

Change each of the following percents to a decimal equivalent. (4 each)

		Decimal			Decimal
1. 30% =	1. _____		2. 5% =	2. _____	
3. 75% =	3. _____		4. 6% =	4. _____	
5. $28\frac{1}{4}$% =	5. _____		6. 325% =	6. _____	
7. 750% =	7. _____		8. 25% =	8. _____	
9. $\frac{1}{2}$% =	9. _____		10. 80% =	10. _____	
11. $7\frac{1}{2}$% =	11. _____		12. 145% =	12. _____	
13. 0.7% =	13. _____		14. $83\frac{1}{3}$% =	14. _____	
15. 0.45% =	15. _____		16. 32.5% =	16. _____	
17. 0.25% =	17. _____		18. $\frac{7}{8}$% =	18. _____	

Change each of the following to decimal and common fraction equivalents. (4 each answer)

		Decimal Fraction	Common Fraction			Decimal Fraction	Common Fraction
19. 20% =	19. _____ _____			20. 40% =	20. _____ _____		
21. 87.5% =	21. _____ _____			22. 75% =	22. _____ _____		
23. 125% =	23. _____ _____			24. $37\frac{1}{2}$% =	24. _____ _____		
25. $\frac{1}{4}$% =	25. _____ _____			26. 0.65% =	26. _____ _____		
27. 750% =	27. _____ _____			28. 55% =	28. _____ _____		
29. $33\frac{1}{3}$% =	29. _____ _____			30. 8.75% =	30. _____ _____		
31. 0.8% =	31. _____ _____			32. $62\frac{1}{2}$% =	32. _____ _____		
33. $8\frac{1}{3}$% =	33. _____ _____			34. $\frac{3}{8}$% =	34. _____ _____		

© Copyright South-Western Publishing Co.

LESSON 63
Changing Decimals and Fractions to Percents

1. To change a decimal fraction to a percent, move the decimal point two places to the right and attach the percent sign.

   Example: Change (a) 0.2, (b) 1.5, and (c) $0.07\frac{1}{2}$ to percents.

   (a) $0.2 = 20\%$

   (b) $1.5 = 150\%$

   (c) $0.07\frac{1}{2} = 7\frac{1}{2}\%$ or 7.5%

2. To change a common fraction to a percent, first change the fraction to a decimal and then to a percent.

   Example: Change (a) $\frac{1}{20}$, (b) $\frac{1}{8}$, and (c) $1\frac{1}{4}$ to percents.

   (a) $\frac{1}{20} = 1 \div 20 = 0.05 = 5\%$

   (b) $\frac{1}{8} = 1 \div 8 = 0.125 = 12.5\%$ or $12\frac{1}{2}\%$

   (c) $1\frac{1}{4} = 1 + (1 \div 4) = 1 + 0.25 = 1.25 = 125\%$

EXERCISE 63/Basic Time–10 Minutes

Estimated time to obtain a basic score of 100.

Change each of the following to a percent. (4 each)

1. $0.35 =$ 1. _____
3. $0.3 =$ 3. _____
5. $2.35 =$ 5. _____
7. $1.67 =$ 7. _____
9. $0.80 =$ 9. _____
11. $0.8 =$ 11. _____
13. $0.09 =$ 13. _____
15. $0.06\frac{1}{4} =$ 15. _____
17. $6.5\% =$ 17. _____
19. $0.00\frac{1}{5} =$ 19. _____

2. $0.60 =$ 2. _____
4. $0.65 =$ 4. _____
6. $0.4 =$ 6. _____
8. $1.37\frac{1}{2} =$ 8. _____
10. $0.00\frac{1}{4} =$ 10. _____
12. $1.10 =$ 12. _____
14. $0.9 =$ 14. _____
16. $0.315 =$ 16. _____
18. $0.0075 =$ 18. _____
20. $10.5 =$ 20. _____

Change each of the following to a percent correct to the nearest tenth of a percent. (6 each)

21. $\frac{1}{2} =$ 21. _____
23. $\frac{2}{5} =$ 23. _____
25. $\frac{5}{8} =$ 25. _____
27. $\frac{7}{8} =$ 27. _____
29. $2\frac{2}{3} =$ 29. _____
31. $\frac{7}{20} =$ 31. _____
33. $7\frac{3}{5} =$ 33. _____
35. $\frac{7}{12} =$ 35. _____
37. $6\frac{1}{2} =$ 37. _____
39. $\frac{5}{12} =$ 39. _____

22. $\frac{1}{4} =$ 22. _____
24. $\frac{1}{6} =$ 24. _____
26. $\frac{1}{50} =$ 26. _____
28. $4\frac{3}{4} =$ 28. _____
30. $\frac{1}{3} =$ 30. _____
32. $\frac{1}{160} =$ 32. _____
34. $\frac{3}{8} =$ 34. _____
36. $\frac{1}{16} =$ 36. _____
38. $\frac{15}{16} =$ 38. _____
40. $5\frac{1}{8} =$ 40. _____

© Copyright South-Western Publishing Co.

TEST 63

Name _____ Total Score _____
Date _____ Basic Score __100__
Basic Time–1 Minute Improvement Score _____

Supply the equivalent common fractions, decimal fractions, and precents in the table below. Show the common fractions in lowest terms and the decimal fractions as hundredths. (4 each)

Use this space for computations	NO.	COMMON FRACTIONS	DECIMAL FRACTIONS	PERCENTS
	1.	$\frac{1}{2}$		
	2.		1.25	
	3.	$\frac{1}{6}$		
	4.	$\frac{3}{8}$		
	5.		0.10	
	6.		1.5	
	7.	$2\frac{3}{4}$		
	8.		$1.33\frac{1}{3}$	
	9.		$0.83\frac{1}{3}$	
	10.	$\frac{7}{8}$		
	11.		4.75	
	12.		$0.01\frac{1}{2}$	
	13.	$\frac{17}{50}$		
	14.		$0.00\frac{1}{4}$	
	15.	$\frac{1}{12}$		
	16.	$\frac{2}{3}$		
	17.		0.03	
	18.			35%
	19.	$\frac{1}{8}$		
	20.		$0.58\frac{1}{3}$	
	21.			$62\frac{1}{2}\%$
	22.	$\frac{4}{5}$		
	23.		$0.00\frac{1}{2}$	
	24.			$6\frac{1}{4}\%$
	25.	$1\frac{1}{10}$		

LESSON 64: Estimation with Percents

As you learned in Exercise 62, *percent* means *hundredths*. Percents offer a way to write decimal fractions in which the denominator is understood to be 100. That is, 1% means 0.01 or $\frac{1}{100}$ and 25% = $\frac{25}{100}$ = $\frac{1}{4}$. Completing this exercise will enable you to better understand the relative sizes of percents.

EXERCISE 64/Basic Time—8 Minutes

Estimated time to obtain a basic score of 100.

Choose one of the three choices in Problems 1-12. Write your answers in the appropriate spaces on the right. (8 each)

1. 48.26% is near $\frac{1}{4}$, $\frac{1}{2}$, or $\frac{3}{4}$. 1. _____
2. 27.32% is near $\frac{1}{4}$, $\frac{1}{2}$, or $\frac{3}{4}$. 2. _____
3. 72.82% is near $\frac{1}{4}$, $\frac{1}{2}$, or $\frac{3}{4}$. 3. _____
4. 9.31% is near $\frac{1}{100}$, $\frac{1}{10}$, or 1. 4. _____
5. 97.67% is near $\frac{1}{100}$, $\frac{1}{10}$, or 1. 5. _____
6. 0.89% is near $\frac{1}{100}$, $\frac{1}{10}$, or 1. 6. _____
7. 19.25% is near $\frac{1}{3}$, $\frac{1}{4}$, or $\frac{1}{5}$. 7. _____
8. 12.7% is near $\frac{1}{4}$, $\frac{1}{8}$, or $\frac{1}{12}$. 8. _____
9. 81.3% is near $\frac{7}{8}$, $\frac{4}{5}$, or $\frac{3}{4}$. 9. _____
10. 34.15% is near $\frac{1}{3}$, $\frac{1}{4}$, or $\frac{1}{5}$. 10. _____
11. 86.09% is near $\frac{3}{4}$, $\frac{4}{5}$, or $\frac{7}{8}$. 11. _____
12. 13.12% is near $\frac{1}{8}$, $\frac{1}{5}$, or $\frac{1}{4}$. 12. _____

Don't compute the exact answer. In the appropriate space on the right, write an estimate for each problem. (8 each)

13. 9.84% × 724 =   13. _____
14. 26.37% × 80 =   14. _____
15. 0.95% × 625 =   15. _____
16. 96.84% × 4,230 =   16. _____
17. 49.32% × 5,670 =   17. _____
18. 76.07% × 428 =   18. _____
19. 21.04% × 5,760 =   19. _____
20. 79.36% × 475 =   20. _____
21. 12.4% × 7,423 =   21. _____
22. 32.65% × 8,130 =   22. _____
23. 87.75% × 7,263 =   23. _____
24. 26% × 312 =   24. _____
25. 48% × 682 =   25. _____

© Copyright South-Western Publishing Co.

TEST 64

Name _____   Total Score _____
Date _____   Basic Score __100__
Basic Time–7 Minutes   Improvement Score _____

Choose one of the three choices in Problems 1-12. Write your answers in the appropriate spaces on the right. (8 each)

1. 98.75% is near $\frac{1}{100}$, $\frac{1}{10}$, or 1.   1. _____

2. 0.93% is near $\frac{1}{100}$, $\frac{1}{10}$, or 1.   2. _____

3. 9.86% is near $\frac{1}{100}$, $\frac{1}{10}$, or 1.   3. _____

4. 24.55% is near $\frac{1}{4}$, $\frac{1}{2}$, or $\frac{3}{4}$.   4. _____

5. 77.25% is near $\frac{1}{4}$, $\frac{1}{2}$, or $\frac{3}{4}$.   5. _____

6. 51.75% is near $\frac{1}{4}$, $\frac{1}{2}$, or $\frac{3}{4}$.   6. _____

7. 26.13% is near $\frac{1}{3}$, $\frac{1}{4}$, or $\frac{1}{5}$.   7. _____

8. 78.2% is near $\frac{3}{4}$, $\frac{4}{5}$, or $\frac{7}{8}$.   8. _____

9. 11.75% is near $\frac{1}{16}$, $\frac{1}{12}$, or $\frac{1}{8}$.   9. _____

10. 21.05% is near $\frac{1}{3}$, $\frac{1}{4}$, or $\frac{1}{5}$.   10. _____

11. 31.45% is near $\frac{1}{3}$, $\frac{1}{4}$, or $\frac{1}{5}$.   11. _____

12. 19.72% is near $\frac{1}{8}$, $\frac{1}{5}$, or $\frac{1}{4}$.   12. _____

Don't compute the exact answer. In the appropriate space on the right, write an estimate for each answer. (8 each)

13. 73.9% × 880 =   13. _____   14. 13.38% × 720 =   14. _____

15. 9.67% × 900 =   15. _____   16. 0.94% × 8,400 =   16. _____

17. 26.52% × 680 =   17. _____   18. 55.37% × 4,200 =   18. _____

19. 41.14% × 5,500 = 19. _____   20. 34.63% × 480 =   20. _____

21. 73.4% × 9,600 =   21. _____   22. 12.65% × 4,480 =   22. _____

22. 26.95% × 8,280 = 23. _____   24. 48% × 714 =   24. _____

25. 79% × 950 =   25. _____

LESSON 65
Percentage Problems

Percentage is the portion or part that results from the multiplication of a number by a percent. The percent number is called the *rate*, and the number multiplied by the rate is known as the *base*. The relationship of these is shown in the equation P = RB.
A. To find the percentage, simply multiply the given number (the base) by the percent.
B. To find the percent (rate) one part is of another, divide the part by the other number and show the answer as a percent.
C. To find the *base* number when an amount and a percent are given, divide the given amount by the percent.

Example A: What number is 4% of 600?
$P = 4\% \times 600$
$P = 0.04 \times 600$
$P = 24$

Example B: 30 is what percent of 500?
$30 = R \times 500$
$30 \div 500 = R$
$0.06 = R$
$6\% = R$

Example C: 40 is 5% of what number?
$40 = 5\% \times B$
$40 \div 0.05 = B$
$800 = B$

EXERCISE 65/Basic Time – 16 Minutes
Estimated time to obtain a basic score of 100.

(10 each)

1. How much is 25% of 384?
2. What number is 225% of 428?
3. What percent of 360 is 45?
4. What percent of 460 is 690?
5. The number 72 is 5% of what number?
6. The number 63 is 35% of what number?
7. What number is 9% of 5?
8. The number 45 is what percent of 48?
9. The number 836 is what percent of 1,000?
10. 32% of what number is 672?
11. 34% of what number is 272?
12. Find 72% of 307.
13. Find 232% of 106.
14. 95 is what percent of 47.5?
15. What percent of 35 is 7?
16. How much is $\frac{1}{2}$% of 8,915?
17. 70.95 is 215% of what number?
18. What number is 0.05% of 971?
19. What percent of 240 is 960?
20. 48 is what percent of 288?

Name	Total Score
Date	Basic Score 100
Basic Time–18 Minutes	Improvement Score

Solve these problems. Write the answers in the spaces on the right. (10 each)

1. What number is 21% of 6?

2. How much is 4.2% of 243?

3. What percent of 888 is 4.44?

4. What percent of 371 is 927.5?

5. The number 135.8 is 28% of what number?

6. The number 175 is 7% of what number?

7. What number is 21% of 87?

8. The number 30 is what percent of 48?

9. The number 572 is what percent of 2,000?

10. 35% of what number is 542.85?

11. 57% of what number is 1,218.66?

12. Find 125% of 89.

13. Find 48% of 564.

14. 114 is what percent of 475?

15. What percent of 45 is 135?

16. How much is $\frac{1}{4}$% of 5,476?

17. 93 is 75% of what number?

18. A merchant's sales this year are 15% greater than last year. If sales last year were $324,756.80, what are the sales this year?

19. A gain on sales of $25,624 was $8,247. The gain is what percent of sales?

20. A merchant had a gain of 25% on sales. The gain was $36,428. Find the sales.

1. _____
2. _____
3. _____
4. _____
5. _____
6. _____
7. _____
8. _____
9. _____
10. _____
11. _____
12. _____
13. _____
14. _____
15. _____
16. _____
17. _____
18. _____
19. _____
20. _____

LESSON 66: Commission on Sales

Agents who sell or buy goods for clients charge a preset rate (percent) based on the selling or buying price. This charge is called a commission. When an agent sells a client's goods, the client receives the selling price less the agent's commission and selling expenses. The amount received by the client is the net proceeds.

1. To find the commission, multiply the sales amount by the rate (changing the rate to a decimal fraction), as shown in Example A.

2. To find the net proceeds, deduct the commission and all other charges from the sales, as shown in Example B.

Example A
Sales × Rate = Commission
$110 × 0.04 = $4.40

Example B
Sales − {Commission + Trucking + Storage + Freight} = Net Proceeds
$110 − ($4.40 + $5 + $4 + $9) = $87.60

EXERCISE 66/Basic Time—15 Minutes

Estimated time to obtain a basic score of 100.

Find the commission and net proceeds:

	SALES	RATE	COMMISSION	TRUCKING	STORAGE	FREIGHT	NET PROCEEDS	
1.	111.15	4%						(6)
2.	275.00	6%						(6)
3.	612.50	$7\frac{1}{2}$%						(8)
4.	437.25	5%						(6)
5.	368.10	6%						(6)
6.	972.38	10%				15.23		(8)
7.	517.14	$6\frac{1}{2}$%			3.75			(12)
8.	249.35	4%		7.50				(10)
9.	1,028.60	5%			4.50			(10)
10.	1,250.75	6%				9.38		(10)
11.	765.50	7%		3.75		8.63		(10)
12.	463.75	6%		12.50		7.35		(10)
13.	948.20	4%			5.00	13.78		(10)
14.	828.00	$4\frac{1}{2}$%		4.50	4.75			(12)
15.	367.40	5%			9.60	9.80		(10)
16.	1,941.12	5%		13.50	12.50	20.25		(12)
17.	534.28	6%		5.75	9.75	6.75		(12)
18.	1,250.00	4%		14.25	12.85	19.50		(10)
19.	638.90	7%		2.50	1.10	13.47		(10)
20.	1,571.50	6%		17.00	14.60	11.75		(10)

21. An agent sold a shipment of oranges for $1,847.25. The charges were as follows: $6 for storage, $8.75 for delivery, and $7\frac{1}{2}$% commission for selling. Find the net proceeds. _____ (12)

Name _____ Total Score _____
Date _____ Basic Score 100
Basic Time–12 Minutes   Improvement Score _____

Find the commission and net proceeds:

	SALES		RATE	COMMISSION		TRUCKING		STORAGE		FREIGHT		NET PROCEEDS		
1.	2,375	00	6%			11	25							(12)
2.	1,130	50	5%											(10)
3.	713	25	7%			12	75	3	00	4	63			(14)
4.	347	60	4%			10	75			9	40			(14)
5.	481	85	6%			9	50			12	35			(14)
6.	817	15	5%			3	00	2	50	4	25			(14)
7.	623	50	4%			7	35	9	85	4	00			(14)
8.	1,250	00	6%			15	00			14	82			(12)
9.	913	75	$7\frac{1}{2}$%			8	50			6	38			(20)
10.	237	12	7%			9	25			2	56			(14)

11. An agent sold 1,250 bushels of a commodity at $1.88 a bushel, paid 7¢ a bushel for trucking, $78.75 for freight, kept 6% for a commission, and sent her client a check for the net proceeds. What was the amount of the check?   11. _____ (24)

12. An agent sold 150 cases of canned goods, 24 cans to the case, at 39¢ a can. The shipping charges were $62.50, and the commission was 5% of sales. Find the net proceeds.   12. _____ (20)

13. Find the net proceeds from the sale of 124 tons at $81.75 a ton and a commission of $4\frac{1}{2}$%. The other charges were $52.75.   13. _____ (18)

LESSON 67

Commission on Purchases

When goods are bought by an agent for a client, the amount paid for the goods is the prime cost, and is the base on which the commission is calculated. The client pays the gross cost, which is the sume of the prime cost, the agent's commission, and the buying expenses.

1. To find the commission, multiply the prime cost by the rate, as shown in Example A.
2. To find the gross cost, add the commission and all other charges to the prime cost, as shown in Example B.

Example A Prime Cost × Rate = Commission
$500 × 0.05 = $25

Example B
Prime Cost + {Commission + Trucking + Storage + Freight} = Gross Cost
$500 + ($25 + $13 + $3.50 + $12) = $553.50

EXERCISE 67/Basic Time–15 Minutes
Estimated time to obtain a basic score of 100.

Find the commission and the gross cost:

	PRIME COST	RATE	COMMISSION	TRUCKING	STORAGE	FREIGHT	GROSS COST	
1.	590.80	6%						(8)
2.	575.00	5%						(6)
3.	817.25	7%						(8)
4.	1,175.00	3½%						(8)
5.	243.20	4%						(8)
6.	1,216.75	4%		23.75				(10)
7.	912.65	6%		11.50				(10)
8.	675.00	7½%				11.42		(10)
9.	91.30	5%			5.69			(8)
10.	2,450.00	6%				32.50		(8)
11.	485.75	5%		12.75		8.43		(10)
12.	421.60	7%		9.50	12.25			(10)
13.	530.70	4%			3.50	6.88		(10)
14.	1,061.50	6%		15.50		19.32		(10)
15.	648.35	7%		12.50		7.35		(10)
16.	1,017.80	4%		13.50	17.50	16.10		(10)
17.	623.50	8%		13.00	7.50	5.76		(10)
18.	893.25	6%		4.80	12.25	5.13		(10)
19.	764.38	5%		5.50	3.75	4.38		(10)
20.	1,275.00	6%		15.75	14.50	11.23		(10)

21. An agent bought 75 bales, averaging 294 pounds each, for a client. The agent paid 7½¢ a pound. His commission was 6½%. The other charges totalled $21.75. Find the gross cost. _____ (16)

TEST 67

Name _____ Total Score _____
Date _____   Basic Score __100__
Basic Time–10 Minutes   Improvement Score _____

Find the commission and the gross cost:

	PRIME COST	RATE	COMMISSION	TRUCKING	STORAGE	FREIGHT	GROSS COST	
1.	2,350 00	4%						(8)
2.	1,875 00	5%		13 75	9 60	12 57		(14)
3.	436 75	7%		12 50				(12)
4.	712 50	6%		5 00	13 75	14 85		(14)
5.	256 80	5%		9 25		10 88		(12)
6.	841 60	4%		10 00				(12)
7.	628 30	6%		6 25	7 00			(12)
8.	325 40	8%		12 50		9 15		(12)
9.	278 65	6%		8 75		2 16		(12)
10.	1,931 20	7%		17 75	10 00	17 13		(14)

An agent bought 1,125 cartons of a commodity at 70¢ a carton, paid $3\frac{1}{2}$¢ a carton for trucking and $29.50 for freight, and charged a commission of 5% for buying. What was the gross cost?   11. _____ (22)

An agent bought 144 boxes of fruit at $6.10 a box on a commission of 6%. Other charges amounted to $9.75. Find the gross cost.   12. _____ (22)

For a client, an agent bought 200 boxes of fruit at $4.85 a box and sold them at $6.75 a box. The commissions charged were 3% for buying and 4% for selling. The cost of handling was $22.75. Find the client's gain.   13. _____ (34)

LESSON 68 Trade Discounts

A trade discount is a deduction from a catalog or list price. When two or more trade discounts are given on one invoice, they may be referred to as a *chain or series of discounts*. Discounts may be computed in the order given, with the first rate based on the list price; the second, on the remainder after the amount of the first discount is deducted from the list price; and the third, on the remainder after the amount of the second discount has been subtracted.

Example A Find the net price for merchandise listed at $500, less trade discounts of 20%, 10%, and 5%:

```
$500   list price
 100   less first discount (20%)
$400   remainder
  40   less second discount (10%)
$360   remainder
  18   less third discount (5%)
$342   net price
```

To find a discount rate that is equal to a series of discounts, consider the total amount of any invoice to be equal to 100% and subtract the discount percents in the manner shown. The remainder, the *net price percent*, subtracted from 100% gives the *equivalent discount rate*.

Example B Find the discount rate that is equal to the series of discounts 20%, 10%, and 5%:

```
100.0%   list price percent
 20.0%   less first discount (20%)
 80.0%   remainder
  8.0%   less second discount (10%)
 72.0%   remainder
  3.6%   less third discount (5%)
 68.4%   net price percent
```

100% − 68.4% = 31.6% equivalent discount rate

EXERCISE 68/Basic Time—20 Minutes

Estimated time to obtain a basic score of 100.

Find the amount of discount and the net price for each of these: (8 each answer)

	LIST PRICE	TRADE DISCOUNTS	AMOUNT OF DISCOUNT	NET PRICE
1.	$600.00	20%, 10%	_____	_____
2.	800.00	25%, 10%	_____	_____
3.	840.00	30%, 25%	_____	_____
4.	750.00	20%, 20%, 10%	_____	_____
5.	979.47	40%, 10%, 10%	_____	_____

Find the equivalent discount rate for each series of discounts: (6 each)

	SERIES OF DISCOUNTS	EQUIVALENT DISCOUNT RATE		SERIES OF DISCOUNTS	EQUIVALENT DISCOUNT RATE
6.	20%, 10%	_____	7.	15%, 10%	_____
8.	30%, 25%	_____	9.	25%, 10%	_____
10.	50%, 50%	_____	11.	10%, 10%, 5%	_____
12.	20%, 10%, 5%	_____	13.	20%, 20%, 10%	_____
14.	30%, 20%, 10%	_____	15.	40%, 10%, 10%	_____

Complete this table: (4 each answer)

	LIST PRICE	TRADE DISCOUNTS	EQUIVALENT DISCOUNT RATE	AMOUNT OF DISCOUNT	NET PRICE
16.	$1,200.00	25%, 20%	_____	_____	_____
17.	3,610.00	40%, 10%	_____	_____	_____
18.	2,672.00	25%, 10%, 5%	_____	_____	_____
19.	944.80	20%, 30%, 40%	_____	_____	_____
20.	1,863.40	50%, 25%, 25%	_____	_____	_____

Name _____ Total Score _____
Date _____ Basic Score __100__
Basic Time–10 Minutes Improvement Score _____

Solve these problems. Write the answers in the spaces on the right.
(20 each problem)

1. A catalog lists tires at $96 each. A discount sheet sent later to the customers of this company announces a trade discount of $18\frac{3}{4}$% on these tires. Determine the net price a customer would pay for each tire ordered.
   1._____

2. An invoice totaling $2,357.10 for tools received from a tool company showed trade discounts of 20% and 10%. Determine the net price to the buyer.
   2._____

3. Terra Manufacturing Company allows trade discounts of $14\frac{2}{7}$% and $12\frac{1}{2}$% to retailers. What is the net price a retailer should pay for an invoice totaling $434?
   3._____

4. A used car priced at $3,925 was sold at a 15% discount in a state that has a 7% sales tax. How much did the buyer pay for the car?
   4._____

5. Westgate Company allows discounts of 25% and 10%, A company salesperson sold merchandise totaling $772 and, in error, gave an equivalent discount of 35%. How much was the salesperson's error?
   5._____

6. How much should a retailer pay the manufacturer for 60 blouses listed at $45 each if trade discounts of 25% and 20% are allowed?
   6._____

7. What are the equivalent discount rates for (a) discounts of 25%, 20%, and 10% and (b) discounts of 15%, 10%, and 5%?
   7a._____
   7b._____

8. Valley Camera Company's catalog lists Model B55 at $120. (a) What is the equivalent discount to Olympic Camera Shop if it is allowed discounts of $33\frac{1}{3}$% and 20%? (b) What is the net price to the shop for camera Model B55?
   8a._____
   8b._____

9. A customer saved $123 by buying a lawn mower at a sale. If the dealer allowed a discount of 25% on this model, what was the original price of the lawn mower?
   9._____

10. Which is the better buy for the customer: (a) a desk listed at $260 with discounts of 25% and 15% or (b) an equal quality desk listed at $250 with discounts of 20%, 15%, and 5%? How much better?
    10._____
    _____

152 © Copyright South-Western Publishing Co.

LESSON 69

Determining Due Dates

The due date for an invoice or loan is the date on which payment is due. When the time is expressed in days, the due date is exactly that number of days after the date of the loan. When the time is expressed as a number of months or years, the due date falls on the same date in the appropriate number of subsequent months or years. If the month does not contain that date, the last day of the month is taken. That is, a one-month note dated May 31 falls due on June 30. When time is given in days, there are two methods for determining due dates.

The exact-time method requires that you know the exact number of days in each month and whether February falls in a leap year. February in a leap year (any year date divisible by 4) has 29 days.

A leap year contains 366 days; a regular year, 365 days. See Example A.

The compound method of finding the number of days between two dates is based on the business assumption that each year is composed of 12 months of 30 days each for a total of 360 days. This method involves the subtraction of the years, months, and days. Like units are arranged in columns, with the earlier date placed below the later date. As in the subtraction of any numbers, it may be necessary to "borrow" a unit from the left. See Example B.

Example A Use the exact-time method to find the number of days from December 18, 1993, to March 14, 1995.

Dec. 18, 1993, to Dec. 18, 1994	365
Dec. 18, 1994, to Dec. 31, 1994	13
January, 1995	31
February, 1995	28
To March 14, 1995	14
Total	451 days

Example B Use the compound method to find the number of days from December 18, 1993, to March 14, 1995.

Year	Month	Day
		14
6	~~2~~	44
1995	~~3~~	~~14~~
1995	12	18
1 yr	2 mo	26 days

360 days + 60 days + 26 days = 446 days

EXERCISE 69/Basic Time—12 Minutes

Estimated time to obtain a basic score of 100.

Find the number of days between the dates shown using (a) the exact-time method and (b) the compound method: (10 each answer)

                                                                                     a   b

1. January 15, 1994, to December 1, 1994
2. February 12, 1994, to January 16, 1995
3. November 3, 1994, to May 30, 1995
4. June 2, 1995, to May 6, 1996
5. November 6, 1994, to March 5, 1995
6. December 8, 1995, to July 20, 1996
7. October 1, 1998, to April 4, 1999

Using the exact-time method, find the due date for each of the following: (10 each)

	DATE OF NOTE	TIME	DUE DATE
8.	March 17, 1995	30 days	8.
9.	February 26, 1994	60 days	9.
10.	August 30, 1993	120 days	10.
11.	July 8, 1994	3 months	11.
12.	September 3, 1995	6 months	12.
13.	January 25, 1997	2 years	13.

© Copyright South-Western Publishing Co.

TEST 69

Name _____ Total Score _____
Date _____   Basic Score 100
Basic Time–15 Minutes   Improvement Score _____

Find the number of days between the dates shown using (a) the exact-time method and (b) the compound method: (8 each, Column a; 6 each, Column b)

		a	b
1.	February 25, 1995, to October 5, 1995	1. _____	_____
2.	March 28, 1995, to March 1, 1996	2. _____	_____
3.	May 14, 1997, to October 25, 1997	3. _____	_____
4.	July 6, 1995 to January 30, 1996	4. _____	_____
5.	September 9, 1994, to October 1, 1996	5. _____	_____
6.	August 29, 1994, to April 1, 1995	6. _____	_____
7.	January 16, 1996, to May 4, 1996	7. _____	_____
8.	February 9, 1995, to January 1, 1996	8. _____	_____
9.	April 15, 1994, to February 18, 1995	9. _____	_____
10.	June 30, 1999, to December 4, 1999	10. _____	_____

Using the exact-time method, find the due date for each of the following: (6 each)

	DATE OF INVOICE OR NOTE	TIME	DUE DATE
11.	April 7, 1993	30 days	11. _____
12.	February 5, 1996	45 days	12. _____
13.	January 13, 1994	90 days	13. _____
14.	September 24, 1995	150 days	14. _____
15.	December 26, 1997	240 days	15. _____
16.	March 31, 2000	3 months	16. _____
17.	October 21, 1998	6 months	17. _____
18.	April 30, 1997	9 months	18. _____
19.	November 19, 1997	$1\frac{1}{2}$ years	19. _____
20.	May 18, 1999	$2\frac{1}{4}$ years	20. _____

LESSON 70 Cash Discounts

As a means of encouraging their customers to make prompt and early payment, sellers often grant special discounts in addition to trade discounts. Such discounts granted for early payment are called cash discounts. They are part of the credit terms and appear on the invoice.

For example, terms of *2/10, n/30* (2% discount in 10 days, net in 30) mean the credit period is 30 days, but the buyer may deduct 2% if the invoice is paid within 10 days from date of invoice. If the invoice is not paid in 30 days, it is past due and may be subject to an interest charge or late fee. Due dates for cash discounts are calculated by the exact-time method shown in Exercise 69.

Example What amount is needed to pay a $200 invoice, terms 2/10, n/30, if it is paid (a) within 10 days from date of invoice or (b) within 11 to 30 days from date of invoice?

(a) $200 amount after trade discounts
    　 4 less cash discount of 2%
    $196 amount to be paid

(b) The cash discount period has expired. Therefore, $200 (the amount of the invoice) is to be paid.

EXERCISE 70/Basic Time–12 Minutes
Estimated time to obtain a basic score of 100.

Find the cash discount that should be taken and the net amount for each of the following: (6 each, Cash Discount column; 4 each, Net Amount column)

	AMOUNT OF INVOICE	TERMS		CASH DISCOUNT	NET AMOUNT
1.	$ 760.00	2/10, n/30	1.		
2.	480.00	1/15, n/60	2.		
3.	707.00	3/20, n/60	3.		
4.	1,259.00	1/10, n/30	4.		
5.	3,490.00	3/30, n/90	5.		
6.	946.50	4/10, 2/20, n/60	6.		
7.	1,040.65	5/10, 2/30, n/60	7.		
8.	3,170.75	2/10, 1/20, n/30	8.		
9.	873.51	6/10, 4/30, n/90	9.		
10.	824.69	5/15, 2/30, n/60	10.		

	DATE OF INVOICE	AMOUNT OF INVOICE	TERMS	DATE PAID		CASH DISCOUNT	NET AMOUNT
11.	Mar. 13	$ 980.00	2/10, n/30	Mar. 22	11.		
12.	Mar. 15	660.00	3/10, 2/30, n/60	Mar. 24	12.		
13.	Mar. 18	875.80	2/10, 1/30, n/60	Apr. 11	13.		
14.	Apr. 18	1,197.00	4/10, 2/30, n/60	Apr. 28	14.		
15.	June 28	1,937.50	5/15, 2/30, n/60	July 13	15.		
16.	July 27	2,557.95	6/10, 2/40, n/90	Aug. 6	16.		
17.	Aug. 10	3,680.75	7/15, 4/45, n/90	Sept. 24	17.		
18.	Sept. 23	1,019.86	2/10, 1/20, n/60	Oct. 14	18.		
19.	Nov. 2	4,186.13	8/10, 4/30, n/90	Dec. 28	19.		
20.	Dec. 16	5,190.60	4/10, 2/30, n/60	Jan. 10	20.		

Name _____ Total Score _____

Date _____ Basic Score __100__

Basic Time–10 Minutes Improvement Score _____

Solve these problems. Write the answers in the spaces on the right.
(20 each problem)

1. An invoice for $896.49 is dated March 26 and has terms of 4/10, n/30. Find (a) the final date on which the cash discount may be taken and (b) the amount necessary to pay this invoice in full if the cash discount is earned.

    1a. _____

    1b. _____

2. An invoice totaling $475, terms 3/15, n/60, is dated March 30. Find (a) the final date on which the cash discount may be taken and (b) the amount that will pay the invoice in full if the cash discount is earned.

    2a. _____

    2b. _____

3. An invoice totaling $2,347.25 is dated May 22 and has terms of 3/10, 2/20, 1/40, n/70. How much would pay this invoice in full on June 11?

    3. _____

4. An invoice for $2,165 is dated December 15 and has terms of 4/30, 2/60, n/90. Find the amount necessary to pay this invoice in full on January 15.

    4. _____

5. On May 28 the R & B Company purchased raw materials invoiced at $3,558 with terms of 5/10, 2/30, n/60. The company paid the invoice in full, less the discount, on June 8. What was the amount of the payment?

    5. _____

6. The regular terms offered by the Cheviot Company are 3/10, 1/30, n/60. On June 17 Pamela Carr purchased merchandise worth $446.75 from the company. On June 27 she purchased merchandise worth $129.80. On July 7 she paid both invoices. What was the amount of the total payment?

    6. _____

7. On January 24 a merchant bought goods invoiced at $654.32, terms 2/10, n/30. The merchant returned part of the goods and received $65 credit for the return. On February 4 a check for the balance due was sent. What was the amount of the check?

    7. _____

8. An invoice for goods amounting to $1,441 was subject to discounts of 25%, 15%, and 10% with terms of 3/15, n/60. How much was the net cash price if paid within the discount period?

    8. _____

9. How much was paid for merchandise invoiced at $1,267 and subject to discounts of 25% and 10%, terms 2/10, n/30, if purchased on August 27 and paid for on September 6?

    9. _____

10. What is the least amount a merchant would have to pay for 25 lamps listed at $120 each if trade discounts of $33\frac{1}{3}$%, 10%, and 10% are allowed and if the terms are 4/10, 2/30, n/90?

    10. _____

LESSON 71 Discounts on Invoices

Invoices from wholesale houses often contain a variety of articles. Some of the articles are sold at certain rates of discount, while others are sold at different rates. Those articles having the same rates of trade discount are grouped together in order to save discount calculations. Total these net prices to obtain the net amount of the bill.

If a cash discount is allowed for payment within a certain time, figure the cash discount on the net amount of the bill and deduct it from the net amount. Determine the due date by the exact-time method.

EXERCISE 71/Basic Time—12 Minutes

Estimated time to obtain a basic score of 100.

Figure the amount that will pay the following bill on January 31, 19--:

```
                                                    INVOICE
            VIKING SUPPLIES, INC.                   NO. 6406
       3460 Brady Street      Fort Worth, TX 76109-2525
                  Tel: (817) 379-0914


SOLD TO    Ms. Parma Samad              DATE    January 5, 19--
           138 E. Berry Street
           Fort Worth, TX 76110-4030    TERMS   2/30, n/60
```

QUANTITY	DESCRIPTION	UNIT PRICE	AMOUNT	NET AMOUNT
3	Calculator (pocket)	$12.50		
4	Stapler	8.95	_____	
	Subtotal		_____(8)	
	Less 10% and 5%		_____(30)	_____(4)
6 doz.	Writing pads	4.00		
10 doz.	Scratch pads	3.75		
5 doz.	Pencils (mechanical)	5.75		
3 doz.	Markers	6.65		
5 doz.	Pens (ballpoint)	7.10	_____	
	Subtotal		_____(20)	
	Less 20% and 10%		_____(20)	_____(4)
325 sheets	Cardboard (per 100)	4.24		
275 sheets	Pressboard (per doz.)	3.80		
500 sheets	Paper (per 100)	5.75		
65 sheets	Cover (plastic)	0.49		
2	Binder	4.75		
26	Folder	0.50		
18	Lamp (desk)	29.50		
24	Tray (plastic)	1.75		
24	Labels	0.25	_____	
	Subtotal		_____(72)	
	Less 25% and 20%		_____(16)	_____(4)
	Total of invoice			_____(10)
	Less cash discount			_____(8)
	Net amount paid			_____(4)

TEST 71

Name _____ Total Score _____
Date _____ Basic Score __100__
Basic Time–12 Minutes Improvement Score _____

Figure the amount that will pay the following bill on December 28, 19--:

DATE	December 18, 19--		INVOICE NO. 6613	
TERMS	2/10, n/30		**PS** PAPERWORKS SHOP	
SOLD TO	Mr. Arturo Herrera 1200 Madison Avenue Syracuse, NY 13210-5325		450 LAFAYETTE ROAD SYRACUSE, NY 13205-6070 AREA CODE (315) 650-6000	

QUANTITY	DESCRIPTION	UNIT PRICE	AMOUNT	NET AMOUNT
6 doz.	Markers	$7.50		
24	Pens (fine point)	6.00		
18 doz.	Pens (ballpoint)	7.25		
30 doz.	Pencils (lead)	1.35		
12	Pen & pencil sets	5.80		
	Subtotal		_____(16)	
	Less 20% and 5%		_____(22)	_____(4)
12	Ribbons (calculator)	3.45		
15	Ribbons (printer)	6.20		
	Subtotal		_____(8)	
	Less 15% and 10%		_____(26)	_____(4)
25	Pads (plastic)	0.50		
50	Pads (paper)	1.25		
	Subtotal		_____(8)	
	Less 10% and 10%		_____(18)	_____(4)
10	Notebook	2.75		
8	Binder (data)	4.50		
6	Binder (ring)	6.50		
12	Binder (pressboard)	5.40		
6	Binder (heavy duty)	8.25		
12	Labels	1.75		
36	Cover (report)	3.35		
	Subtotal		_____(30)	
	Less 20%, 10%, and 10%		_____(34)	_____(4)
	Total of invoice			_____(10)
	Less cash discount			_____(8)
	Net amount paid			_____(4)

158 © Copyright South-Western Publishing Co.

LESSON 72
Simple Interest Formula

Interest is an amount paid for the use of money. A bank discount is interest collected in advance by a bank at the time money is borrowed from it.

In order to find the ordinary simple interest (or bank discount) on any sum of money for any length of time at any rate of interest, you may use the formula *Interest = Principal × Rate × Time in Years* or abbreviated as I = PRT. See Example A.

Example A Find the interest on $900 at 6% for two years (I = PRT).
I = $900 × 0.06 × 2 = $108

Interest is charged on an annual basis. When the length of time of a loan is stated in months, the number of months is placed over 12. See Example B.

Example B Find the interest on $900 at 6% for 3 months (I = PRT).
I = $900 × 0.06 × $\frac{3}{12}$ = $13.50

For the problems in this book, place the number of days over 360 when the time is given in days. (See Example C.) Using 360 days in the formula yields *ordinary interest*. Using 365 days (366 for leap years) gives *exact interest*.

Example C Find the ordinary interest on $900 at 6% for 90 days (I = PRT).
I = $900 × 0.06 × $\frac{90}{360}$ = $13.50

EXERCISE 72/Basic Time–18 Minutes
Estimated time to obtain a basic score of 100.

Find the ordinary simple interest for each of the following:

	PRINCIPAL	RATE	YEARS	INTEREST
1.	$725	6%	1	____(10)
2.		7%	2	____(10)
3.		8%	5	____(12)

	PRINCIPAL	RATE	DAYS	INTEREST
7.	$3,330	10%	60	____(10)
8.		9%	80	____(10)
9.		12%	36	____(10)

	PRINCIPAL	RATE	MONTHS	INTEREST
4.	$3,500	6%	3	____(14)
5.		12%	5	____(12)
6.		10%	8	____(20)

	PRINCIPAL	RATE	DAYS	INTEREST
10.	$925	11%	18	____(8)
11.		15%	53	____(16)
12.		9%	57	____(18)

13. What amount of interest would be earned on this note as of June 16, 1994? (Calculate the exact number of days through June 16; then use the 360-day business year for the interest calculation.) _____(30)

```
$1,250.00              Pittsburgh, Pa.           March 1, 19 94
Six months             AFTER DATE       I        PROMISE TO PAY TO
THE ORDER OF   Charles M. Anderson
                       PAYABLE AT First National Bank
Twelve Hundred Fifty and 00/100- - - - - - - - - - - - - - - - - - - - DOLLARS
VALUE RECEIVED WITH INTEREST AT 7%
No. 111  Due _____
                                      L.D. Frye
```

14. What amount of interest would be paid when this note is due? (Use the business-year basis for interest calculation.) _____(20)

```
$1,750.00              Chicago, Ill.             December 24, 19 96
Six months             AFTER DATE       I        PROMISE TO PAY TO
THE ORDER OF   C.H. Kramer
                       PAYABLE AT First National Bank
One Thousand Seven Hundred Fifty and 00/100- - - - - - - - - - - - - DOLLARS
VALUE RECEIVED WITH INTEREST AT 8%
No. 322  Due _____
                                      D.M. Ritter
```

Name _____ Total Score _____

Date _____ Basic Score __100__

Basic Time–10 Minutes Improvement Score _____

1. Find the interest on the following note to maturity. (Use the business-year basis for the computation.)

   1. _____ (24)

   ```
   $1,625.00           Atlanta, Georgia      August 5, 19 96
   Forty-five Days      AFTER DATE    I     PROMISE TO PAY TO
   THE ORDER OF Friedman Tailors
                        PAYABLE AT Merchants Bank
   One Thousand Six Hundred Twenty-five and 00/100 -------- DOLLARS
   VALUE RECEIVED WITH INTEREST AT 12%
   No. 222  Due _____
                                      R.K. Ballister
   ```

On the above note, what would the interest have been:

2. At 7%? 2. _____ (8)
3. At 8%? 3. _____ (14)
4. At 10%? 4. _____ (14)
5. At 14%? 5. _____ (14)

6. Find the interest on the following note to maturity. (Calculate the exact number of days to maturity. Then use the business-year basis for the interest calculation.)

   6. _____ (50)

   ```
   $2,760.00           Atlanta, Georgia      February 15, 19 96
   _____           AFTER DATE    I     PROMISE TO PAY TO
   THE ORDER OF The First National Bank
                        PAYABLE AT Merchants Bank
   Two Thousand Seven Hundred Sixty and 00/100 -------- DOLLARS
   VALUE RECEIVED WITH INTEREST AT 9%
   No. 445  Due June 30, 1996
                                      Mary Hart
   ```

On the above note, what would the interest have been:

7. At 10%? 7. _____ (18)
8. At $14\frac{1}{2}$%? 8. _____ (18)
9. At $7\frac{1}{2}$%? 9. _____ (24)
10. At 18%? 10. _____ (16)

LESSON 73: Compound Interest

Compound interest is interest on the sum of an original principal plus its accumulated interest. For example, money placed in a savings account at a bank will earn interest at a given rate. The bank adds the interest earned to the account periodically (monthly, or quarterly).

The added interest increases the balance in the account, which earns interest during the next period. Thus, interest added at the end of one period becomes principal in the next period. The sum of the original principal and its compound interest is the compound amount.

Example A Judy Kepler deposited $1,000 in a fund that pays 10% compounded annually. Find (a) the compound amount and (b) the compound interest at the end of two years.

Solution:
Original principal	$1,000
Interest for 1st period	100
Compound amount	$1,100
Interest for 2d period	110
(a) Compound amount	$1,210

(b) $1,210 – $1,000 = $210 compound interest

Example B Jack Pence deposited $1,000 in a fund that pays 10% compounded semiannually. Find the compound amount at the end of two years.

Solution:
Original principal	$1,000.00
Interest for 1st period	50.00
Compound amount	$1,050.00
Interest for 2d period	52.50
Compound amount	$1,102.50
Interest for 3d period	55.13
Compound amount	$1,157.63
Interest for 4th period	57.88
Compound amount	$1,215.51

EXERCISE 73 / Basic Time – 24 Minutes

Estimated time to obtain a basic score of 100.

For each problem, find (a) the compound amount and (b) the compound interest: (10 each answer)

	PRINCIPAL	RATE	TIME	COMPOUNDED	COMPOUND AMOUNT	COMPOUND INTEREST
1.	$900	15%	3 years	annually	_____	_____
2.	1,500	10%	4 years	annually	_____	_____
3.	8,000	14%	2 years	semiannually	_____	_____
4.	23,000	9%	$2\frac{1}{2}$ years	semiannually	_____	_____
5.	70,000	16%	1 year	quarterly	_____	_____
6.	64,000	12%	$\frac{1}{4}$ year	monthly	_____	_____

Solve these problems. Place each answer in the appropriate space on the right. (10 each answer)

7. Leonard Drake deposited $30,000 in an account that pays 8% compounded quarterly. Find the compound amount at the end of one year.

7. _____

8. Matilda Johnson invested $25,000 in a certificate that paid 15% compounded monthly. How much interest did this certificate earn in four months?

8. _____

9. Four years ago a trust fund of $50,000 was invested at 12.5% compounded annually. Find the amount in the fund now.

9. _____

10. Jesse Williams inherited $75,000 on the day he became twenty-one years old. The money was placed in a trust fund that earned interest at 15% compounded semiannually. What amount was in the fund on his twenty-third birthday?

10. _____

11. Fifteen months ago the original amount in a fund was $15,000. The interest rate is 12%. Find (a) the simple interest and (b) the interest compounded quarterly.

11a. _____
11b. _____

12. The sum of $46,000 was invested at 18% compounded monthly. (a) How much was in the fund at the end of three months? (b) How much more was earned with compound interest than would have been earned with simple interest?

12a. _____
12b. _____

Name _____ Total Score _____

Date _____ Basic Score __100__

Basic Time–21 Minutes   Improvement Score _____

For each problem, find (a) the compound amount and (b) the compound interest: (10 each answer)

	PRINCIPAL	RATE	TIME	COMPOUNDED	COMPOUND AMOUNT	COMPOUND INTEREST
1.	$800	9%	3 years	annually	_____	_____
2.	2,400	8%	1 year	quarterly	_____	_____
3.	7,000	14%	$1\frac{1}{2}$ years	semiannually	_____	_____
4.	25,000	15%	$\frac{1}{4}$ year	monthly	_____	_____
5.	90,000	11%	2 years	semiannually	_____	_____
6.	36,000	18%	$\frac{1}{3}$ year	monthly	_____	_____

Solve these problems. Write each answer in the appropriate space on the right. (10 each answer)

7. Wilma Conci deposited $40,000 in an account that pays 9% compounded monthly. How much compound interest did this account earn in three months?

7. _____

8. Charles Struble invested $32,000 in a certificate that paid 14% compounded quarterly. Find the compound amount at the end of nine months.

8. _____

9. Two years ago a trust fund of $60,000 was invested at 13% compounded semiannually. Find the amount in the fund now.

9. _____

10. Lucille Odum inherited $85,000 on the day she became eighteen years old. The money was placed in a trust fund that earned interest at 14% compounded annually. What amount was in the fund on her twenty-first birthday?

10. _____

11. Five months ago the amount in a fund was $43,000. The interest rate is 12%. Find (a) the simple interest and (b) the interest compounded monthly.

11a. _____

11b. _____

12. The sum of $27,000 was invested at 16% compounded quarterly. (a) How much was in the fund at the end of one year? (b) How much more was earned with compound interest than would have been earned with simple interest?

12a. _____

12b. _____

162 © Copyright South-Western Publishing Co.

5 Cumulative Review

Name _____

Write each answer in the appropriate space on the right.

Change each of the following percents to decimal notation.

1. $\frac{1}{5}$%

2. 0.7%

Change each of the following percents to a common fraction or mixed number in lowest terms.

3. 4.25%

4. $93\frac{3}{4}$%

Change each of the following to a percent.

5. 4.75

6. $0.00\frac{1}{2}$

Solve these problems.

7. How much is 45% more than $160?

8. How much is 7% less than $480?

9. About $16\frac{2}{3}$% of a novely item sold by a company are returned to the company by its customers. During the past year total sales of this item amounted to $557,749.80. How much were the actual sales?

10. Betty Walters' salary is $2,880 a month. This is 20% more than Jane Burkitt earns. How much is Jane's monthly salary?

11. $22.80 is what percent of $80?

12. What percent of $390 is $198.90?

13. A young man who earns $405 a week gave his sister $135 as her birthday present. What percent of his weekly earnings did he give her?

14. A radio that cost a dealer $48 was sold for $64. The difference between the cost and the selling price is what percent of the cost?

15. $714 is $62\frac{1}{2}$% of what amount?

16. 25% of how many dollars is $47?

17. Peggy Kendall invested 55% of her cash in real estate and had $45,900 left. How much cash did she have before she bought the real estate?

18. A real estate agent sold an apartment building for $1,750,000. How much did the seller receive if the agent deducted 4% commission and $526 authorized expenses.

1. _____
2. _____
3. _____
4. _____
5. _____
6. _____
7. _____
8. _____
9. _____
10. _____
11. _____
12. _____
13. _____
14. _____
15. _____
16. _____
17. _____
18. _____

© Copyright South-Western Publishing Co.

5 Cumulative Review

19. If an agent charged $119.50 for selling $4,780 worth of commmodities, what rate of commission was charged?

    19. _____

20. If an agent received $330 for selling goods at $7\frac{1}{2}$% commission, what was the selling price of the goods?

    20. _____

21. An egg rancher sent 35 cases of eggs (30 dozen to a case) to a commission merchant who sold the eggs at 89¢ a dozen. The charges were: freight, $15.35; commission, 7%; and other expenses, $12.83. How much should the rancher receive?

    21. _____

22. A commission merchant received an order for 620 boxes of fruit to be bought on a commission of $3\frac{1}{2}$%. There were 300 boxes bought at $5.50 a box; 200 at $5.60 a box, and 120 at $5.75 a box. Also, $78.40 was paid for freight. Find the gross cost.

    22. _____

23. There are how many days between the dates of July 26 and November 2 of the same year?

    23. _____

24. $\underline{\ ?\ } \times 10\% \times \dfrac{5}{12} = \$125$.

    24. _____

25. $\$5{,}000 \times \underline{\ ?\ } \times \dfrac{45}{360} = \$56.25$

    25. _____

26. $\$8{,}000 \times 18\% \times \dfrac{?}{360} = \$160$

    26. _____

27. Alpine Products Company charges 18% interest on overdue bills for the exact number of days that each bill is overdue. How much is needed to pay in full a bill for $450 that was due on June 9 but not paid until September 10?

    27. _____

28. A loan produced $75 interest at 12% for 45 days. Find the principal of the loan.

    28. _____

29. At what interest rate will $675 yield $13.50 interest in 3 months?

    29. _____

30. How many days are needed for $486 to yield $6.48 interest at 8%?

    30. _____

31. In compound interest, the interest on $1,000 at a given rate will be larger in 2 years if the compounding is done (quarterly/monthly) __?__.

    31. _____

32. If interest is compounded semiannually, the interest is computed __?__ times a year.

    32. _____

33. Two years ago Angelica Quezada invested $2,000 at 6% interest compounded semiannually. The amount in the fund now is __?__.

    33. _____

34. A discount that is given to encourage customers to pay promptly is a __?__ discount.

    34. _____

35. Retailers who do not pay an invoice within the cash-discount period (do/do not) __?__ forfeit the cash discount.

    35. _____

36. The equivalent discount rate for the series 30% and 10% is __?__.

    36. _____

5 Cumulative Review

37. The net-price percent for the series 25% and 20% is _?_.

38. Retailers who do not pay off invoices within the cash-discount period (do/do not) _?_ forfeit their right to deduct the trade discounts offered on the invoice.

39. A dealer received an invoice for $2,132.50 of merchandise that is subject to trade discounts of 20% and 15%. The net price to the dealer is _?_.

40. An invoice for $632 has terms of 2/10, n/30 and is dated March 23. The amount to be paid on April 2 is _?_.

41. How much is 32% of $92.10?

42. What is 16% of 41?

43. What percent of 132 is 231?

44. 125 is 25% of what number?

45. Find the interest on $1,215 at $10\frac{1}{2}$% for 1 year.

46. Find the interest on $675 at 8% for 180 days.

47. Find the interest on $540 at 12% for 90 days.

48. Find the compound interest on $8,000 at 14% compounded quarterly for 1 year.

49. Find the interest at $14\frac{1}{2}$% on $8,450 for 45 days.

50. Find the interest on $896.50 at 15% for 75 days.

51. Find the amount to be paid on an invoice for $340.78, terms 5/10, 3/30, n/60, which was dated September 26 and paid on October 26.

52. Lew was born on April 18, 1970. His brother David was born on June 4, 1973. How much older is Lew than David? *(Use the compound method.)*

53. Amelia Lugo wishes to save $7,500 in one year for her vacation. If her annual salary is $30,000, what percent of her salary should she save for the vacation?

54. An agent sells merchandise on 12% commission. What is the total sales value of the merchandise that must be sold in order to earn $600 in commission?

55. An agent sold 750 crates of produce for a grower at $7.65 each and 157 crates at $7.50 each. The agent paid storage charges of $37.49 and freight charges of $72.89. How much net proceeds were due the grower if the agent's commission rate was $4\frac{1}{2}$%?

56. How much was paid for merchandise purchased on September 23, invoiced at $1,385, subject to trade discounts of 30% and 20%, terms 2/10, n/30, and paid for on October 3?

57. What is 2.3% of 89?

Cumulative Review

58. 95 is 20% of what amount?

59. What percent of $947 is $217.81?

60. What number is 23% of $875?

61. Find the interest on $8,767.75 at $10\frac{1}{2}$% for 240 days.

62. Find the interest at $9\frac{1}{4}$% on $1,750 for 30 days.

63. Find the interest on $1,250 at 15% for 4 months.

64. Find the compound amount on $6,000 at 12% compounded semiannually for 2 years.

65. How much is the compound interest on $9,000 for 3 months at 18% compounded monthly?

66. Find the amount to be paid on an invoice for $917.23, terms 4/10, 2/30, n/60, dated on July 16, and paid on August 16.

67. A student traveled in Europe from June 16 to October 8. The student was in Europe for exactly how many days?

68. An automobile traveling at 60 miles per hour moves at a speed of 5,280 feet a minute. How far will it move in 45 seconds?

69. A commission merchant bought and shipped to a Denver company 375 boxes of a product at $6.95 a box and 250 boxes at $5.80 a box. Charges were $47.25 for storage and $108.59 for freight. If the rate of commission was $6\frac{1}{2}$%, how much gross cost was paid by the Denver company?

70. A merchant purchased 75 radios listed at $48 each. Special trade discounts were as follows: $16\frac{2}{3}$%, $12\frac{1}{2}$%, and 20%; terms 5/10, 2/20, n/60. What is the least that must be paid?

58. _____
59. _____
60. _____
61. _____
62. _____
63. _____
64. _____
65. _____
66. _____
67. _____
68. _____
69. _____
70. _____

LESSON 74: U. S. Measurements

In this exercise, common measurements are given. Review thoroughly the list of equivalent values presented in the right-hand column below.

The problems to be solved require that measurements in one denomination be changed to equivalent higher or lower denominations.

You may need scratch paper for figuring. Write each answer in the space allotted, and show any remainder as a fractional part in lowest terms.

Example: How many minutes are there in 260 seconds?
Solution: 60 seconds = 1 minute
Therefore: 260 seconds ÷ 60 seconds = $4\frac{20}{60}$ minutes = $4\frac{1}{3}$ minutes

EXERCISE 74/Basic Time—7 Minutes

Estimated time to obtain a basic score of 100.

Supply the required answers as indicated: (8 each)

#	Problem	#	Answer unit	Equivalents
1.	768 inches =	1. _____	feet	12 in. = 1 ft
2.	102 feet =	2. _____	yards	3 ft = 1 yd
3.	7,920 feet =	3. _____	miles	5,280 ft = 1 mi
4.	430 seconds =	4. _____	minutes	60 sec = 1 min
5.	843 minutes =	5. _____	hours	60 min = 1 hr
6.	1,020 hours =	6. _____	days	24 hours = 1 day
7.	4,165 days =	7. _____	weeks	7 days = 1 wk
8.	1,095 days =	8. _____	years	365 days = 1 yr
9.	792 months =	9. _____	years	12 mo = 1 yr
10.	754 weeks =	10. _____	years	52 wk = 1 yr
11.	864 square inches =	11. _____	square feet	144 sq in. = 1 sq ft
12.	801 square feet =	12. _____	square yards	9 sq ft = 1 sq yd
13.	5,120 acres =	13. _____	square miles	640 acres = 1 sq mi
14.	15,552 cubic inches =	14. _____	cubic feet	1,728 cu in. = 1 cu ft
15.	351 cubic feet =	15. _____	cubic yards	27 cu ft = 1 cu yd
16.	1,076 fluid ounces =	16. _____	quarts	32 fl oz = 1 qt
17.	483 pints =	17. _____	quarts	2 pt = 1 qt
18.	528 quarts =	18. _____	gallons	4 qt = 1 gal
19.	2,079 cubic inches =	19. _____	gallons	231 cu in = 1 gal
20.	224 articles =	20. _____	dozen	12 articles = 1 doz
21.	468 dozen =	21. _____	gross	12 doz = 1 gross (gr)
22.	1,152 articles =	22. _____	gross	144 articles = 1 gr
23.	1,232 ounces =	23. _____	pounds	16 oz = 1 pound (lb)
24.	9,000 pounds =	24. _____	tons	2,000 lb = 1 ton (T)
25.	60 mills =	25. _____	cents	10 mills = 1 cent

Name _____ Total Score _____

Date _____ Basic Score __100__

Basic Time—9 Minutes Improvement Score _____

Solve these problems. Write the answers in the spaces on the right. (20 each)

1. A stationer sold 7 gross of pencils at 2 pencils for 5 cents. What was the total amount received for the pencils?   1._____

2. A wheel, 15 feet in circumference, makes how many revolutions in going one mile?   2._____

3. How many days are there in 90 hours?   3._____

4. How many square yards of carpet will cover a floor containing 117 square feet?   4._____

5. A bridge abutment contains 3,861 cubic feet. How many cubic yards are there in it?   5._____

6. How many pieces of tile of 24 square inches each will cover 336 square feet of floor?   6._____

7. How many yards of tape are required for 28 pieces of 27 inches each?   7._____

8. How many posts, 3 yards apart, will be needed for a fence around an estate 3 miles around?   8._____

9. How many hours are there in 205,200 seconds?   9._____

10. How many weeks are there in 525 days?   10._____

LESSON 75

Denominate Numbers

Numbers such as 6, $3\frac{1}{2}$, and 7.5 that do not refer to objects are abstract numbers. Numbers that do refer to objects, such as 6 pencils, $3\frac{1}{2}$ pies, and 8.5 gallons, are concrete numbers. Concrete numbers that are expressed in terms of standard units of measure are denominate numbers. For example, 5 dollars, 6 weeks, $7\frac{1}{4}$ pounds, and 8.5 gallons are denominate numbers; they identify the measures of quantity.

To add or subtract denominate numbers, arrange the numbers in columns with the lowest denomination on the right so that like denominations will be added or subtracted.

Example A   Add: 6 yd   1 ft   8 in.
                        3 yd   2 ft   7 in.
                    10 yd   1 ft   3 in.

Solution A Add 8 in. and 7 in. obtaining 15 in., or 1 ft and 3 in. Write 3 under the inches column and carry 1 to the next column to the left, the total of which becomes 4 ft. As 4 ft make 1 yd and 1 ft, write 1 under the feet column and carry 1 to the next column, the total of which becomes 10 yd.

Example B   Subtract: 4̶   5̶ hr   2̶9̶ 30 min   8̶0̶ 2̶0̶ sec
                             3 hr   45 min   30 sec
                             1 hr   44 min   50 sec

Solution B. Borrow 1 min from the 30 min to make 29 min and 80 sec. Subtract 30 sec from 80 sec and write the difference, 50 sec, in the "seconds" column. Borrow 1 hr from the 5 hr to give 4 hr 89 min. Subtract 45 min from 89 min and 3 hr from 4 hr. Write the answers in the appropriate columns.

EXERCISE 75/Basic Time—7 Minutes

Estimated time to obtain a basic score of 100.

Add: (20 each)

1. 6 days   4 hr   15 min
    5 "   3 "   24 "
    7 "   11 "   17 "
    2 "   10 "   5 "

2. $7   8 ct   4 mills
    4   37 "   9 "
    12   84 "   6 "
    3   26 "   8 "

3. 16 hr   11 min   20 sec
    12 "   29 "   16 "
     3 "   15 "   25 "
     6 "   18 "   18 "

4. 15 gr   10 doz   4 articles
     2 "   11 "   8 "
     3 "   9 "   6 "
    12 "   8 "   7 "

Subtract: (20 each)

5. 6 gal   1 qt   0 pt
    3 "   3 "   1 "

6. 76 cu yd   2 cu ft   100 cu in.
    75 " "   8 " "   121 " "

7. 7 yr   9 wk   6 da
    3 "   12 "   9 "

8. 16 sq yd   5 sq ft   12 sq in.
    15 " "   8 " "   27 " "

9. 10 yd   1 ft   10 in.
     4 "   2 "   11 "

10. 24 hr   1 min   5 sec
    17 "   5 "   7 "

Name _____ Total Score _____
Date _____ Basic Score __100__
Basic Time–5 Minutes Improvement Score _____

Add: (20 each)

1. 15 gr   10 doz   9 articles
    12 "   11 "   10 "
    11 "   5 "   9 "
     5 "   9 "   11 "
    ─────   ─────   ─────────

2. 12 hr   14 min   14 sec
     7 "   10 "   38 "
     3 "   18 "   20 "
     1 "   19 "   23 "
    ─────   ─────   ─────

3. 11 yr   10 mo
    10 "   8 "
     9 "   8 "
     7 "   4 "
    ─────   ─────

4.   1 yd   2 ft   11 in.
     2 "   1 "   8 "
                2 "   7 "
     6 "   10 "
    ─────   ─────   ─────

5.   2 sq yd   8 sq ft   120 sq in.
     3 " "   5 " "   82 " "
     6 " "   7 " "   57 " "
     1 " "   2 " "   13 " "
    ──────── ──────── ──────────

6.   9 gal   3 qt   1 pt
    25 "   2 "
    17 "   1 "   1 "
     7 "   1 "
    ─────   ─────   ─────

Subtract: (20 each)

7. 20 lb   14 oz
    12 "   15 "
    ─────   ─────

8.   7 yr   10 wk   14 da
     2 "   15 "   18 "
    ───── ─────   ─────

9. 1,927 days 4 hr   16 min
    1,921 "   5 "   24 "
    ────────── ───── ──────

10. 17 gr   4 doz   6 articles
     11 "   9 "   11 "
    ─────   ─────   ─────────

LESSON 76
Vocabulary of Metric System of Measurement

The metric system of measuring length, capacity, weight, and area is used in science and industry in the United States and is the official system that is used in most countries of the world. The names of the units of measure commonly used in the metric system are as follows:

Name	Pronunciation	Meaning
meter (m)	meeter	unit of length
liter (L)	leeter	unit of capacity
gram (g)	gram	unit of weight

Prefixes are added to the foregoing names to indicate 10, 100, or 1 000 (and other powers of 10) times the unit of measure being used. For example, dekaliter means 10 liters, hectogram means 100 grams, and kilometer means 1 000 meters. Likewise, other prefixes may be appended to each name of a unit of measure to show 0.1, 0.01, or 0.001 part of. Notice that decigram means 0.1 gram; centimeter, 0.01 meter; and milliliter, 0.001 liter. A space, rather than a comma is used to separate large numbers into groups of three. A zero is always used before a decimal point.

Prefix	Meaning	Numerical Value
kilo (k)	one thousand times	1 000
hecto (h)	one hundred times	100
deka (da)	ten times	10
	base unit (meter, liter, gram)	1
deci (d)	one tenth part of	0.1
centi (c)	one hundredth part of	0.01
milli (m)	one thousandth part of	0.001

EXERCISE 76/Basic Time–2 Minutes
Estimated time to obtain a basic score of 100.

In the appropriate space on the right, place the name or prefix that is asked for in each question. Do not abbreviate. (20 each)

1. What name is applied to the standard unit of measure for length in the metric system? 1._____

2. In the metric system of measurement, what is the name of the standard unit of measure for weight? 2._____

3. In the metric system, what is the name of the standard unit of measure for capacity? 3._____

4. What is used instead of the comma to separate large numbers into groups of three in the metric system? 4._____

5. What prefix is used in the metric system to mean one thousandth part of? 5._____

6. In the metric system, what is the prefix that is used to mean ten times? 6._____

7. What prefix is used in the metric system to mean one tenth part of? 7._____

8. In the metric system, what prefix is used to mean one hundred times? 8._____

9. In the metric system, what prefix is used to mean one hundredth part of? 9._____

10. What is the prefix that is used in the metric system to mean one thousand times? 10._____

Name _____ Total Score _____
Date _____ Basic Score 100
Basic Time–2 Minutes Improvement Score _____

Place each answer in the appropriate space on the right. Do not abbreviate. (20 each)

1. In the metric system of measurement, what is the name of the standard unit of measure for weight?
   1. _____

2. What name is applied to the standard unit of measure for capacity in the metric system?
   2. _____

3. In the metric system, what is the name of the standard unit of measure for length?
   3. _____

4. What prefix is used in the metric system to mean one hundredth part of?
   4. _____

5. What prefix is used in the metric system to mean one thousandth part of?
   5. _____

6. In the metric system, what prefix is used to mean one hundred times?
   6. _____

7. What prefix is used in the metric system to mean one tenth part of?
   7. _____

8. What is the prefix that is used in the metric system of measurement to mean one thousand times?
   8. _____

9. In the metric system, what prefix is used to mean 10 times?
   9. _____

10. Write the number ten thousand as it appears in the metric system.
    10. _____

LESSON 77: Metric Measurements

Changing from a metric unit of measurement to a larger or smaller metric unit may be done by moving the decimal point to the left or right. Using a chart like the one shown in the examples simplifies the procedure. The prefixes in the chart are arranged in order of size.

To change a metric unit to a larger metric unit, move the decimal point one place to the left for each higher unit.

Example A shows that deka is *four* places to the *left* of milli. To change the milli units to deka units, the point is moved four places to the left

To change a metric unit to a smaller metric unit, move the decimal point one place to the right for each lower unit.

Example B shows that centi is *two* places to the *right* of the base unit. Therefore, the decimal point is moved two places to the right.

Example A Change 8 000 millimeters to dekameters.
Base
kilo hecto deka **Unit** deci centi milli
Solution: 8 000 mm = 8 000 = 0.8 dam

Example B Change 700 grams to centigrams.
Base
kilo hecto deka **Unit** deci centi milli
Solution: 700 g = 700.00 = 70 000 cg

meters (m)		liters (L)		grams (g)	
kilometer (km) =	1 000 m	kiloliter (kL) =	1 000 L	kilogram (kg) =	1 000 g
hectometer (hm) =	100 m	hectoliter (hL) =	100 L	hectogram (hg) =	100 g
dekameter (dam) =	10 m	dekaliter (daL) =	10 L	dekagram (dag) =	10 g
meter (m) =	1 m	liter (L) =	1 L	gram (g) =	1 g
decimeter (dm) =	0.1 m	deciliter (dL) =	0.1 L	decigram (dg) =	0.1 g
centimeter (cm) =	0.01 m	centiliter (cL) =	0.01 L	centigram (cg) =	0.01 g
millimeter (mm) =	0.001 m	milliliter (mL) =	0.001 L	milligram (mg) =	0.001 g

EXERCISE 77/Basic Time–12 Minutes

Estimated time to obtain a basic score of 100.

Write the full name for each of the following symbols: (10 each)

1. 3 kg   1. _____   2. 2 mL   2. _____
3. 1 cm   3. _____   4. 7 dm   4. _____
5. 1 g   5. _____   6. 6 L   6. _____

Change the metric values below as indicated: (10 each)

SMALLER TO LARGER

7. 40 mm =   7. _____ cm
8. 59 g =   8. _____ dag
9. 400 cg =   9. _____ dg
10. 20 dm =   10. _____ m
11. 30 mm =   11. _____ dm
12. 300 L =   12. _____ kL
13. 200 g =   13. _____ kg

LARGER TO SMALLER

14. 2 cL =   14. _____ mL
15. 15 kg =   15. _____ hg
16. 8 L =   16. _____ dL
17. 6 dm =   17. _____ cm
18. 4 g =   18. _____ cg
19. 2 km =   19. _____ dam
20. 7 kg =   20. _____ g

TEST 77

Name _____ Total Score _____
Date _____ Basic Score 100
Basic Time–12 Minutes Improvement Score _____

Write the full name for each of the following symbols: (10 each)

1. 1 mm
2. 4 dag
3. 3 cL
4. 3 hg
5. 2 km
6. 5 dL

1. _____
2. _____
3. _____
4. _____
5. _____
6. _____

Change the metric values below as indicated: (10 each)

7. 4 mm =
8. 300 g =
9. 5 L =
10. 20 km =
11. 8 dL =
12. 60 g =
13. 4 kL =
14. 0.001 m =
15. 0.9 g =
16. 90 L =
17. 7 daL =
18. 600 m =
19. 3 dg =
20. 80 hL =

7. _____ cm
8. _____ kg
9. _____ cL
10. _____ dam
11. _____ mL
12. _____ dag
13. _____ L
14. _____ dm
15. _____ dg
16. _____ hL
17. _____ hL
18. _____ km
19. _____ cg
20. _____ kL

LESSON 78 — Metric Denominates

Metric denominates are added, subtracted, multiplied, and divided just like any other values, but they must always be expressed in equivalent units before any calculations are made.

In Examples A and B the metric denominates are changed to equivalent units before the addition or subtraction. If decimals are involved, the values must be aligned on the decimal points.

In Examples C and D the multiplication and division of the metric denominates proceed in the usual way. The answers are expressed in the metric unit of measure that is used.

Example A
Solution: Add 6.5 meters and 35 centimeters.
Change the 35 centimeters to 0.35 meter. Then add the meters:
```
  6.5  m
 +0.35 m
  6.85 m
```

Example B Subtract 125 millimeters from 75 centimeters.

Solution Change 125 millimeters to 12.5 centimeters. Then subtract:
```
 75.0 cm
-12.5 cm
 62.5 cm
```

Example C Multiply 12.3 meters by 5.

Solution:
```
 12.3 m
 x  5
 61.5 m
```

Example D Divide 620 millimeters by 4.

Solution:
```
     155 mm
   4)620 mm
```

EXERCISE 78 / Basic Time – 6 Minutes
Estimated time to obtain a basic score of 100.

Add: (12 each)

1. 70 mL + 625 mL + 325 mL = _____ mL
2. 40 g + 0.9 kg = _____ g
3. 30 km + 5 m = _____ km
4. 45.2 L + 3.75 L + 7.945 L = _____ L
5. 6 g + 20 dg = _____ g

Multiply: (8 each)

11. 345 L × 20 = _____ L
12. 0.36 kL × 8 = _____ kL
13. 0.75 g × 1.5 = _____ g
14. 0.6 mm × 268 = _____ mm
15. 55 kg × 2.5 = _____ kg

Subtract: (12 each)

6. 5 g − 1 mg = _____ g
7. 5 m − 15 mm = _____ m
8. 85 cm − 150 mm = _____ cm
9. 6 dL − 60 mL = _____ dL
10. 7 g − 600 mg = _____ g

Divide: (8 each)

16. 39 L ÷ 3 = _____ L
17. 625.5 kg ÷ 5 = _____ kg
18. 1 004 m ÷ 8 = _____ m
19. 876 mg ÷ 24 = _____ mg
20. 361.2 mL ÷ 14 = _____ mL

TEST 78

Name _____ Total Score _____
Date _____ Basic Score __100__
Basic Time–12 Minutes   Improvement Score _____

Add: (6 each)

1. 6 dg + 25 dg + 153 dg = _____ dg
2. 82.25 g + 46.037 g = _____ g
3. 75 cL + 4.5 L = _____ cL
4. 350 cm + 8.75 m = _____ m
5. 28 daL + 30 L = _____ daL
6. 6 m + 48 mm = _____ mm
7. 54 g + 1.2 hg = _____ hg
8. 2.4 L + 10.8 dL = _____ L
9. 38 dag + 648 g = _____ dag
10. 5.3 km + 728 m = _____ km

Subtract: (6 each)

11. 34 m – 47 dm = _____ m
12. 125 g – 8 dag = _____ g
13. 968 L – 6 hL = _____ L
14. 27 dL – 8.1 mL = _____ dL
15. 4.3 g – 9.7 mg = _____ mg
16. 218.7 m – 1 hm = _____ hm
17. 65.6 m – 465 cm = _____ cm
18. 19.6 hg – 683 g = _____ g
19. 59.04 kL – 54.9 hL = _____ kL
20. 1771 cm – 1.25 m = _____ m

Multiply: (4 each)

21. 150 g × 6 = _____ g
22. 43 m × 8 = _____ m
23. 48.6 L × 9 = _____ L
24. 74 hL × 16 = _____ hL
25. 13 cm × 25 = _____ cm
26. 15.3 × 39 mg = _____ mg
27. 9.4 × 4.86 dL = _____ dL
28. 2.5 × 36 km = _____ km
29. 64 × 6.25 dag = _____ dag
30. 32 × 12.5 cL = _____ cL

Divide: (4 each)

31. 318 L ÷ 4 = _____ L
32. 864 m ÷ 8 = _____ m
33. 6 459.6 g ÷ 12 = _____ g
34. 756 dg ÷ 5 = _____ dg
35. 6 598.9 mm ÷ 11 = _____ mm
36. 39 L ÷ 6 = _____ L
37. 91.492 dam ÷ 8.9 = _____ dam
38. 8.19 kg ÷ 3.5 = _____ kg
39. 93.6 dL ÷ 7.8 = _____ dL
40. 520.56 hg ÷ 9.64 = _____ hg

LESSON 79
Converting U. S. Measurements to Metric Measurements

United States units of measure may be changed to metric units by multiplying the number of U. S. units by the *metric equivalent* of 1 U. S. unit. The procedure may be done in two steps:
1. Determine the metric equivalent of 1 unit of the U. S. measurement that is to be changed. (The equivalent metric unit is called the conversion factor.)
2. Multiply the conversion factor by the quantity to be changed.

Sometimes the decimal point in the conversion factor that is found in the table must be moved to the right or to the left. Notice in Example B that 1 inch equals 2.540 centimeters. As a centimeter is 0.01 meter, 2.540 centimeters equal 0.0254 meter (the conversion factor).

Example A   3 quarts = __?__ liters
Solution: 1 quart = 0.9463 liter (Table I)
Therefore: 3 qt. = 3 × 0.9463 L = 2.8389 L

Example B   50 inches = __?__ meters
Solution: 1 inch = 2.540 cm = 0.0254 meter
Therefore: 50 in = 50 × 0.0254 = 1.27 m

EXERCISE 79/Basic Time – 20 Minutes
Estimated time to obtain a basic score of 100.

Use Table I on page 185 to solve these problems. Where applicable, round the answer to the nearest ten-thousandth. Place each answer in the appropriate space on the right. (10 each)

1. 5 feet = _____ meters
2. 20 ounces = _____ grams
3. 12 liquid quarts = _____ liters
4. 32 acres = _____ hectares
5. 75 inches = _____ centimeters
6. 3 pounds = _____ grams
7. 56 fluid ounces = _____ milliliters
8. 8 cubic feet = _____ cubic meters
9. 30 miles = _____ kilometers
10. 9 liquid pints = _____ liters
11. 7 short tons = _____ metric tons
12. 90 fluid ounces = _____ liters
13. 25 gallons = _____ dekaliters
14. 15 inches = _____ meters
15. 50 pounds = _____ kilograms
16. 45 long tons = _____ metric tons
17. 35 inches = _____ decimeters
18. 9 ounces = _____ centigrams
19. 27 cubic inches = _____ cubic centimeters
20. 60 square feet = _____ square meters

TEST 79

Name _____ Total Score _____

Date _____ Basic Score 100

Basic Time–18 Minutes   Improvement Score _____

Use Table I on page 185 to solve these problems. Where applicable, round the answer to the nearest ten-thousandth. Place each answer in the appropriate space on the right. (10 each)

1. 50 fluid ounces = _____ milliliters

2. 12 pounds = _____ grams

3. 60 inches = _____ centimeters

4. 8 acres = _____ hectares

5. 15 liquid quarts = _____ liters

6. 18 ounces = _____ grams

7. 27 feet = _____ meters

8. 30 yards = _____ meters

9. 17 liquid pints = _____ liters

10. 56 miles = _____ kilometers

11. 40 fluid ounces = _____ liters

12. 9 short tons = _____ metric tons

13. 4 ounces = _____ centigrams

14. 70 inches = _____ decimeters

15. 37 long tons = _____ metric tons

16. 58 pounds = _____ kilograms

17. 26 inches = _____ meters

18. 45 pounds = _____ dekagrams

19. 32 square feet = _____ square meters

20. 68 cubic inches = _____ cubic centimeters

178   © Copyright South-Western Publishing Co.

LESSON 80

Converting Metric Measurements to U.S. Measurements

Metric units of measure may be changed to U. S. units by multiplying the number of metric units by the *U. S. equivalent* of 1 metric unit. Follow these two steps:

1. Find the U. S. equivalent of 1 unit of the metric measurement that is to be changed. (The U. S. equivalent unit is the conversion factor.)
2. Multiply the quantity to be changed by the conversion factor.

On occasion, the desired conversion factor may be obtained by moving the decimal point in some other factor found in the table. Observe that in Example B, the conversion factor (3.28084 ft) for one meter is obtained by moving the decimal point in the equivalent in feet of one dekameter (32.8084 ft).

Example A	Five liters equal how many liquid quarts?
Solution:	1 liter = 1.0567 quarts (Table II)
Therefore:	5 L = 5 × 1.0567 qt
	= 5.2835 qt

Example B	How many feet are there in 6 meters?
Solution:	1 dekameter = 32.8084 ft (Table II)
	1 meter = 3.28084 ft
Therefore:	6 m = 6 × 3.28084 ft = 19.6850 ft

EXERCISE 80/Basic Time–20 Minutes

Estimated time to obtain a basic score of 100.

Use Table II on page 186 to solve these problems. Where applicable, round the answer to the nearest ten-thousandth. Place each answer in the appropriate space on the right. (10 each)

1. Seven liters equal how many liquid quarts?
2. Five grams equal how many ounces?
3. Twelve square meters equal how many square yards?
4. How many inches are there in 4 meters?
5. How many pounds are there in 265 grams?
6. How many liquid pints are there in 9 liters?
7. There are how many acres in 4 hectares?
8. There are how many miles in 25 kilometers?
9. There are how many feet in 40 meters?
10. Twenty hectares equal how many acres?
11. Thirty square centimeters equal how many square inches?
12. Fourteen square meters equal how many square feet?
13. How many short tons are there in 18 metric tons?
14. How many fluid ounces are there in 44 milliliters?
15. There are how many fluid ounces in 75 centiliters?
16. There are how many yards in 130 meters?
17. Nineteen kilograms equal how many pounds?
18. Sixty dekameters equal how many feet?
19. How many square miles are there in 25 square kilometers?
20. How many ounces are there in 60 decigrams?

Name _____ Total Score _____
Date _____ Basic Score 100
Basic Time–8 Minutes Improvement Score _____

Use Table II on page 186 to solve these problems. Where applicable, round the answer to the nearest ten-thousandth. Place each answer in the appropriate space on the right. (10 each)

1. There are how many ounces in 16 grams? 1. _____
2. There are how many acres in 15 hectares? 2. _____
3. Eleven meters equal how many inches? 3. _____
4. Seventy-one grams equal how many pounds? 4. _____
5. Fifteen liters equal how many gallons? 5. _____
6. How many square inches are there in 26 square centimeters? 6. _____
7. How many square yards are there in 5 square meters? 7. _____
8. How many liquid pints are there in 3 liters? 8. _____
9. How many meters are there in 12 yards? 9. _____
10. How many feet are there in 25 meters? 10. _____
11. Forty kilometers equal how many miles? 11. _____
12. Twenty liters equal how many liquid quarts? 12. _____
13. How many fluid ounces are there in 35 centiliters? 13. _____
14. How many square feet are there in 24 square meters? 14. _____
15. There are how many U. S. tons in 19 metric tons? 15. _____
16. There are how many fluid ounces in 57 milliliters? 16. _____
17. How many inches are there in 625 centimeters? 17. _____
18. How many pounds are there in 32 dekagrams? 18. _____
19. Thirty-five decimeters equal how many inches? 19. _____
20. Fifty square kilometers equal how many square miles? 20. _____

6 Cumulative Review

Name _____

Write each answer in the appropriate space on the right.

1. In the addition of measurements, the largest units of measure should be added (first/last) _?_.
   1._____

2. The prefix *kilo* has a numeric value of _?_
   2._____

3. Metric symbols always should be shown in the (singular/plural) _?_ form.
   3._____

4. To express a U. S. measurement in a smaller unit, the measurement is (multiplied/divided) _?_ by the number of smaller units needed to make one of the larger units.
   4._____

5. In the subtraction of measurements, the smallest units of measure should be subtracted (first/last) _?_.
   5._____

6. The prefix *deka* has a numeric value of _?_.
   6._____

7. The prefix *milli* has a numeric value of _?_.
   7._____

8. The basic metric unit of measure for weight is the _?_.
   8._____

9. The basic unit of measure for capacity is the _?_.
   9._____

10. The prefix *hecto* has a numeric value of _?_.
    10._____

11. The prefix *deci* has a numeric value of _?_.
    11._____

12. One hectare equals how many ares?
    12._____

13. This measurement is expressed (correctly/incorrectly) _?_: The area of the poster is 360 dm^2.
    13._____

14. This measurement is expressed (correctly/incorrectly) _?_: The box is $\frac{1}{2}$ meter tall.
    14._____

15 - 19. Show the metric symbol for each of these.

15. Milligrams   16. Hectoliters

17. Kilograms   18. Cubic meters

19. Square centimeters

15._____
16._____
17._____
18._____
19._____

20. In the metric system, a _?_ instead of a comma is used to separate a numeral into groups of three digits.
    20._____

21. To change a metric value to an equivalent value in a larger metric unit of measure, the decimal point is moved to the (left/right) _?_.
    21._____

22. In the metric system, the _?_ replaces the acre.
    22._____

23 - 26. Convert each of the following.

23. 156 items to dozens.

24. 14 quarts to fluid ounces.

25. 342 pints to gallons and quarts.

26. 24 hours to minutes.

23._____
24._____
25._____
26._____

27. Add: 9 years 5 months 16 days, 7 years 8 months 26 days, and 4 years 3 months 9 days.
    27._____

© Copyright South-Western Publishing Co. **181**

6 Cumulative Review

28. Which is the longest measurement: (a) 17.75 feet, (b) 7.5 yards, or (c) 257 inches?

28. _____

29. A field is 800 yards long and 200 yards wide. How many feet of fencing are needed to enclose this field?

29. _____

30. Subtract 16 gallons 3 quarts 1 pint from 34 gallons 1 quart.

30. _____

31. The bus travels from Fairfield to Greenville in 5 hours 20 minutes. The train makes fewer stops and takes 3 hours 25 minutes. Traveling by train saves how many hours and minutes?

31. _____

32 - 37. Convert each of the following.

32. 6 409 millimeters to dekameters.

32. _____

33. 2 376 kilograms to grams.

33. _____

34. 748 dekameters to centimeters.

34. _____

35. 4 146 hectograms to kilograms.

35. _____

36. 189 deciliters to milliliters.

36. _____

37. 73 152 centiliters to dekaliters.

37. _____

38. Emily Chang plans to fence part of her garden. Each side of the area is 14 meters long. One end is 9.6 meters and the other is 7.75 meters. How many meters of fencing are required?

38. _____

39 - 44. Show the equivalent measurement to the nearest hundredth.

39. 35 meters to feet.

39. _____

40. 48 inches to centimeters.

40. _____

41. 64 ounces to grams.

41. _____

42. 46 kilograms to pounds.

42. _____

43. 32 liters to liquid quarts.

43. _____

44. 58 fluid ounces to milliliters.

44. _____

45. To the nearest inch, how many inches are in 688 centimeters?

45. _____

46. To the nearest tenth, how many liters are in 5 gallons 3 quarts.

46. _____

47. Paula Madrid bought one each of the following lengths of plastic pipe: 3.5 meters, 4 meters 5 centimeters, 2 meters 8 decimeters, and 0.75 meter. She bought how many meters of pipe?

47. _____

48. The distance from an airport near Denver to one near New York is 2 592 kilometers. At an average speed of 960 kilometers per hour, a plane flight takes how many hours and minutes?

48. _____

49. A can of food seasoning that weighs 44 grams sells for $1.54. At this price, 1 kilogram of the seasoning is worth how much?

49. _____

50. An automobile dealer advertises that a certain model car averages 35.2 kilometers per gallon of gasoline. To the nearest tenth, this is a distance of how many miles per gallon?

50. _____

Posttest

Solve these problems. Place each answer in the appropriate space on the right. Show common fractions in lowest terms. (Score 2 points for each correct answer to Problems 1-20 and 3 points for each correct answer to Problems 21-40.)

Add:

1. 465
   203
   197
   842
   357

2. $2.80 + 5.43 + 0.08 + 8.79 =$

3. $\dfrac{5}{6}$
   $\dfrac{7}{12}$

4. $21\dfrac{11}{14}$
   $8\dfrac{3}{7}$

Subtract:

5. 7,600
   2,658

6. $9.35 - 4.765 =$

7. $\dfrac{7}{15} - \dfrac{1}{3} =$

8. $9\dfrac{7}{8}$
   $4\dfrac{1}{4}$

Multiply

9. 468
   503

10. $\dfrac{3}{8} \times 54 =$

11. $\dfrac{8}{9} \times \dfrac{5}{12} =$

12. $18\dfrac{2}{5}$
    $6\dfrac{3}{4}$

13. $70 \times 0.70 =$

14. $3.7 \times 0.8 =$

1. _____
2. _____
3. _____
4. _____
5. _____
6. _____
7. _____
8. _____
9. _____
10. _____
11. _____
12. _____
13. _____
14. _____

Posttest

Divide:

15. $824 \div 8 =$

16. $0.615 \div 0.006 =$

17. $32 \div \frac{3}{8} =$

18. $\frac{5}{9} \div \frac{1}{3} =$

19. $9\frac{5}{8} \div 19 =$

20. $5\frac{3}{4} \div 10\frac{3}{8} =$

Write each of the following as a percent:

21. 0.0375

22. $\frac{3}{5}$

23. 4

Write each of the following as a common fraction:

24. 25%

25. $66\frac{2}{3}\%$

26. $\frac{1}{5}\%$

Write each of the following as a decimal fraction:

27. $1\frac{3}{8}$

28. $62\frac{1}{2}\%$

29. Find the average of these daily attendance numbers:
    1,589   1,578   1,726   1,604   1,723

30. Solve this equation: $3(n + 4) - 5(n - 7) = 235$

Solve each of the following:

31. $87\frac{1}{2}\%$ of $224 = \underline{\ ?\ }$

32. $45 = 5\%$ of $\underline{\ ?\ }$

33. $15 = \underline{\ ?\ }\%$ of 60

34. Add: 6 hr 43 min
            8 hr 35 min

35. 3 250 cm = $\underline{\ ?\ }$ m

36. A student deposited $900 in a savings account that pays 12% interest compounded quarterly. Find the compound amount at the end of one year.

37. How much is the simple interest on $1,500 at 11% for 8 months?

38. On May 9, a dealer listed terms of 2/10, n/30 on an invoice for $1,700 of merchandise. How much should the customer pay on May 18?

39. An agent sold a client's shipment of fruit and collected $3,450. The charges were $97.25 for freight and 5% commission for selling. How much should the agent send to the client?

40. Find the net price for merchandise listed on an invoice at $800 less trade discounts of 15% and 10%.

15. _____
16. _____
17. _____
18. _____
19. _____
20. _____
21. _____
22. _____
23. _____
24. _____
24. _____
26. _____
27. _____
28. _____
29. _____
30. _____
31. _____
32. _____
33. _____
34. _____
35. _____
36. _____
37. _____
38. _____
39. _____
40. _____

TABLE I. U. S. Units of Measure and Their Approximate Metric Equivalents

Cubic
- 1 cubic inch (cu in. or in.³) = 16.3871 cubic centimeters (cm³)
- 1 cubic foot (cu ft or ft³) = 0.0283 cubic meter (m³) = 28.316 liters (L)
- 1 cubic yard (cu yd or yd³) = 0.7646 cubic meter

Dry
- 1 pint (pt) = 33.6003 cubic inches (cu in.) = 0.5506 liter (L)
- 1 quart (qt) = 67.2006 cubic inches = 1.1012 liters
- 1 peck (pk) = 537.605 cubic inches = 8.8095 liters
- 1 bushel (bu) = 2,150.42 cubic inches = 35.2391 liters

Linear
- 1 inch (in.) = 2.540 centimeters (cm)
- 1 foot (ft) = 0.3048 meter (m)
- 1 yard (yd) = 0.9144 meter
- 1 rod (rd) = 5.0292 meters
- 1 mile (mi) = 1.6093 kilometers (km)

Liquid
- 1 fluid dram (fl dr) = 0.2256 cubic inches (cu in.) = 3.6967 milliliters (mL)
- 1 fluid ounce (fl oz.) = 1.8047 cubic inches = 29.5727 milliters
- 1 gill (gi) = 7.2188 cubic inches = 118.2908 milliliters
- 1 pint (pt) = 28.8750 cubic inches = 0.4732 liter (L)
- 1 quart (qt) = 57.7500 cubic inches = 0.9463 liter
- 1 gallon (gal) = 231 cubic inches = 3.7853 liters

Square or Surface
- 1 square inch (sq in. or in.²) = 6.4516 square centimeters (cm²)
- 1 square foot (sq ft or ft²) = 0.0929 square meters (m²)
- 1 square yard (sq yd or yd²) = 0.8361 square meters
- 1 square rod (sq rd or rd²) = 25.2928 square meters
- 1 acre (A) = 0.4047 hectare = 4 046.8564 square meters
- 1 square mile (sq mi or mi²) = 2.5900 square kilometers (km²)

Weight (Avoirdupois)
- 1 grain (gr) = 0.0648 gram (g)
- 1 dram (dr) = 1.7718 grams
- 1 ounce (oz) = 28.3495 grams
- 1 pound (lb) = 453.5924 grams
- 1 short ton (T) = 0.9072 metric ton (t)
- 1 long ton (lt) = 1.01605 metric tons

Metric Units of Measure and Their Approximate U. S. Equivalents

Area

100 square millimeters (mm²)	= 1 sq centimeter (cm²)	= 0.0001 m²	= 0.1550 sq in.
100 square centimeters (cm²)	= 1 sq decimeter (dm²)	= 0.01 m²	= 0.1076 sq ft
100 square decimeters (dm²)	= 1 centare*(ca)	= 1 m²	= 10.7639 sq ft
100 square meters (m²)	= 1 are*(a)	= 100 m²	= 119.5990 sq yd
100 ares (a)	= 1 hectare*(ha)	= 10 000 m²	= 2.4711 acres
100 hectares (ha)	= 1 sq kilometer (km²)	= 1 000 000 m²	= 0.3861 sq mi

*Used in measuring land.

Capacity

		Cubic	Dry	Liquid
	1 milliliter (mL)	= 0.0610 cu in.		= 0.2705 fluidram
10 milliliters (mL)	= 1 centiliter (cL)	= 0.6102 cu in.		= 0.3381 fl oz
10 centiliters (cL)	= 1 deciliter (dL)	= 6.1025 cu in.	= 0.1816 pint	= 0.2113 pint
10 deciliters (dL)	= 1 liter (L)	= 61.0255 cu in.	= 0.9081 quart	= 1.0567 quarts
10 liters (L)	= 1 dekaliter (daL)	= 0.3532 cu ft	= 1.1351 pecks	= 2.6418 gallons
10 dekaliters (daL)	= 1 hectoliter (hL)	= 3.5315 cu ft	= 2.8378 bushels	
10 hectoliters (hL)	= 1 kiloliter (kL)	= 1.3079 cu yd		

Linear

1 millimeter (mm)		= 0.001 meter (m)	= 0.0394 inch
10 millimeters (mm)	= 1 centimeter (cm)	= 0.01 meter	= 0.3937 inch
10 centimeters (cm)	= 1 decimeter (dm)	= 0.1 meter	= 3.9370 inches
10 decimeters (dm)	= 1 meter (m)		= 39.3701 inches
10 meters (m)	= 1 dekameter (dam)	= 10 meters	= 32.8084 feet
10 dekameters (dam)	= 1 hectometer (hm)	= 100 meters	= 328.0840 feet
10 hectometers (hm)	= 1 kilometer (km)	= 1 000 meters	= 0.6214 mile

Weight (Mass)

1 milligram (mg)		= 0.001 gram (g)	= 0.0154 grain
10 milligrams (mg)	= 1 centigram (cg)	= 0.01 gram	= 0.1543 grain
10 centigrams (cg)	= 1 decigram (dg)	= 0.1 gram	= 1.5432 grains
10 decigrams (dg)	= 1 gram (g)		= 0.0353 ounce
10 grams (g)	= 1 dekagram (dag)	= 10 grams	= 0.3527 ounce
10 dekagrams (dag)	= 1 hectogram (hg)	= 100 grams	= 3.5274 ounces
10 hectograms (hg)	= 1 kilogram (kg)	= 1 000 grams	= 2.2046 pounds
1 000 kilograms (kg)	= 1 metric ton (t)	= 1 000 000 grams	= 1.1023 tons

LESSON A1 Calculator Operations

Before starting any problem on a calculator, press the `C` key twice to clear any prior computations. Many pocket calculators also have a *clear-entry* feature. Pressing `CE` clears the number displayed. If there is only one key for both clear and clear entry, one press clears the display and two presses clear the machine.

To add, subtract, multiply, or divide on an algebraic calculator, enter the numbers and signs as you read them from left to right. Signed numbers may be added (but not subtracted, multiplied, or divided) with a pocket calculator.

Add: 18 `+` 32 `+` 25 `=` → 75
Subtract: 745 `−` 396 `=` → 349
Multiply: 53 `×` 72 `=` → 3,816
Divide: 2,170 `÷` 35 `=` → 62

In the addition of signed numbers, 7 `+` (−6) `+` 5 is the same as 7 `−` 6 `+` 5.

EXERCISE A1

Use an 8-digit calculator to solve these problems. (10 each)

Find each total and check accuracy by adding in reverse order. Place each answer in the appropriate space.

1. 9,656 + 7,906 + 5,634 + 975 =
2. 92,707 + 48,374 + 60,943 + 15,637 =
3. 35,608 + 4,738 + 52,678 + 83,949 =

Find the difference and check the accuracy of each answer.

4. 92,163 − 46,065 =
5. 85,700 − 53,748 =
6. 968,648 − 714,147 =

Solve each problem.

7. 84,620 − 24,380 + 9,205 − 46,797 =
8. 99,443 − 27,286 − 36,702 − 44,576 − 21,329 =
9. 38,506 − 89,570 + 76,032 − 5,374 − 90,350 =

Add these signed numbers.

10. −876 + 700 + (−694) + 905 =
11. −453 + 706 + (−847) + (−826) + 954 =
12. −600 + 853 + (−327) + (−879) + 542 =

Multiply. Check accuracy by multiplying the factors in reverse order.

13. 728 × 213 × 52 =
14. 9,295 × 29 × 76 =

Divide. Check accuracy by multiplying the quotient by the divisor.

15. 7,577,265 ÷ 87,095 =
16. 4,714,875 ÷ 75,438 =
17. $\dfrac{4,580,526}{1,739} =$

Solve these problems.

18. 340,575 × 47 ÷ 95 =
19. 3,517,394 ÷ 613 × 87 =
20. $\dfrac{37,362 \times 354}{479} =$

TEST A1

Name _____ Total Score _____
Date _____

Use an 8-digit calculator to solve these problems. (8 each)

Find each total and check accuracy. Place each answer in the appropriate space.

1. 7,856 + 9,706 + 6,454 + 875 =   1. _____
2. 84,707 + 59,437 + 50,493 + 18,736 =   2. _____
3. 46,102 + 3,935 + 69,786 + 74,576 =   3. _____

Find the difference and check the accuracy of each answer.

4. 89,036 − 54,056 =   4. _____
5. 91,500 − 42,437 =   5. _____
6. 985,486 − 674,258 =   6. _____
7. 3,120,958 − 895,076 =   7. _____

Solve each problem.

8. 93,260 − 42,830 + 8,520 − 37,245 =   8. _____
9. 89,343 − 16,682 − 27,602 − 53,765 − 23,192 =   9. _____
10. 47,605 − 76,570 + 89,032 − 6,374 − 50,840 =   10. _____
11. 25,207 + 23,076 − 41,425 + 19,223 − 8,943 =   11. _____

Add these signed numbers.

12. −786 + 900 + (−946) + 509 =   12. _____
13. +943 − 607 + (−384) + (−228) + 459 =   13. _____
14. −800 + 653 + (−732) + (−987) + 254 =   14. _____
15. +910 − 582 + (−832) + 491 + (−706) =   15. _____

Multiply. Check accuracy by multiplying the factors in reverse order.

16. 872 × 321 × 64 =   16. _____
17. 9,592 × 97 × 62 =   17. _____

Divide. Check accuracy by multiplying the quotient by the divisor.

18. 8,525,205 ÷ 76,050 =   18. _____
19. 7,507,773 ÷ 84,357 =   19. _____
20. $\dfrac{5,473,926.4}{1,640}$ =   20. _____
21. $\dfrac{31,231,200}{25,025}$ 21. _____

Solve these problems.

22. 528,093 × 38 ÷ 76 =   22. _____
23. 5,206,590 ÷ 615 × 79 =   23. _____
24. $\dfrac{45,472 \times 246}{580}$ =   24. _____
25. $\dfrac{52,920 \times 630}{945}$ 25. _____

188   © Copyright South-Western Publishing Co.

LESSON A2 Calculator Memory

Review Exercise 32, Order of Operations. Rewriting some equations in *on-line* form helps one to see which operations should be done first. Rewriting the equation:

$$y = \frac{13 \times 2 \times 185}{17 \times 4}$$

in on-line form gives

$y = (13 \times 2 \times 185) \div (17 \times 4)$

or

$y = 13 \times 2 \times 185 \div 17 \div 4$

These keys are common on pocket calculators with a fully addressable memory:

M+ Adds the number in display to the invisible memory register.
M– Subtracts the number in display from the memory register.
MR Recalls and displays the current balance in memory. Does not clear memory.
MC Clears memory balance to zero.

Computing the divisor first and storing it in memory may be more efficient for some division problems.

$n = (75 \times 450) \div (15 \times 25)$

MC 15 ✕ 25 **M+** 75 ✕ 450 ÷ **MR** = → 90

EXERCISE A2

Write the on-line form for each of the following. (8 each)

1. $\dfrac{270 \times 3 \times 7}{15 \times 12} =$

2. $\dfrac{(54 + 8) \times 4}{31} =$

3. $\dfrac{1{,}500 \div (6 + 14)}{34 - 9} =$

4. $\dfrac{(960 - 160) \div 8}{36 + 4} =$

Solve each equation. Use your calculator's memory as needed. (14 each)

5. $a = 18 + 12 \times 36 - 30$

6. $b = 64 \times 30 - 72 \div 24$

7. $c = 240 - \dfrac{180}{45} =$

8. $n = \dfrac{130 \times 104 - 26}{173}$

9. $x = \dfrac{140 \times 224 - 168 \times 112}{64}$

10. $y = \dfrac{224 \div 56 + 32 \times 15}{28 + 16}$

11. $a = (37 \times 50) + (46 \times 35)$

12. $b = (28 \times 65) + (30 \times 54)$

13. $d = (76 \times 48) - (34 \times 42)$

14. $n = (102 \times 37) - (51 \times 68)$

15. $y = (142 \times 129) \div (43 \times 71)$

16. $z = (184 \times 60) \div (46 \times 24)$

TEST A2

Name _____   Total Score _____

Date _____

Write the on-line form for each of the following. (10 each)

1. $\dfrac{45 - 3}{6} =$

2. $\dfrac{9(192 - 7)}{25} =$

3. $\dfrac{360 \times 3 \times 9}{25 \times 12} =$

4. $\dfrac{(54 + 9) \times 4}{21} =$

5. $\dfrac{2880 \div (6 + 12)}{34 + 6} =$

6. $\dfrac{(1020 - 60) \div 6}{36 - 4} =$

Solve each equation. Use an 8-digit calculator. (10 each)

7. $a = 19 + 11 \times 42 - 40$

8. $b = 53 \times 40 - 192 \div 24$

9. $c = 320 - \dfrac{259}{37}$

10. $n = \dfrac{170 \times 102 - 120}{123}$

11. $x = \dfrac{260 \times 142 - 138 \times 123}{160}$

12. $y = \dfrac{736 \div 46 + 52 \times 15}{16 + 24}$

13. $a = (28 \times 67) + (39 \times 53)$

14. $b = (41 \times 75) + (40 \times 64)$

15. $c = (82 \times 43) - (38 \times 39)$

16. $d = (97 \times 62) - (43 \times 72)$

17. $n = (81 \times 27) - (115 \times 58)$

18. $x = (136 \times 48) \div (32 \times 40)$

19. $y = (183 \times 43) \div (61 \times 86) =$

20. $z = (180 \times 56) \div (72 \times 48) =$

LESSON A3

Calculator Fractions and Constants

Calculators process fractions in decimal form. Any common fraction may be changed to decimal form by dividing the numerator by the denominator.

$\frac{1}{2} = 1 \div 2 = 0.5$ a *terminating* decimal fraction.

$\frac{2}{3} = 2 \div 3 = 0.666666...$ a *non-terminating* decimal fraction

For better accuracy, enter as many of the repeating digits in a non-terminating decimal as your calculator will accept

Most pocket calculators are programmed to repeatedly add, subtract, multiply, or divide by an entered number. The repeating number is the *constant*. Enter 4 ➕ 5 🟰; now enter 10 🟰. If the answer in display is 14, the first addend entered is the constant. But if the displayed sum is 15, the second addend entered is the constant. A similar test may be used for multiplication. In subtraction and division, however, the number following the − or ÷ is the constant.

Example: *(First addend is constant; otherwise 156 would be the second addend.)*
156 ➕ 178 🟰 gives 334, then 189 🟰 gives 345, and 196 🟰 gives 352.

Example: *(Second factor is constant; otherwise 7 would be the first factor.)*
58 ✖ 7 🟰 gives 406, then 7.35 🟰 gives 51.45, and 125 🟰 gives 875.

The equals sign may be used to repeatedly add a constant addend or subtract a constant subtrahend from the number in display. Likewise, the equals sign may be used to repeatedly multiply the number in display by a constant factor or divide by a constant divisor.

Example: 7,889 ÷ 7 ÷ 7 ÷ 7
Solution: 7889 ➗ 7 🟰 🟰 🟰 gives 23

EXERCISE A3

Use an 8-digit calculator to solve these problems. (8 each)

For each of these, show the decimal equivalent that should be entered.

1. $\frac{1}{4} =$ 2. $\frac{1}{3} =$ 3. $\frac{1}{9} =$

4. $\frac{3}{5} =$ 5. $\frac{5}{12} =$

Solve these problems. Express your answer as a decimal.

6. Add: $38\frac{1}{4} + 325\frac{1}{3} + 198\frac{7}{8} =$

7. Subtract: $64\frac{1}{8} - 52\frac{2}{3} =$

8. Multiply: $72\frac{3}{8} \times 41\frac{1}{5} =$

9. Divide: $87\frac{1}{16} \div 6\frac{1}{4} =$

Use the appropriate constant for each problem.

10. Add 9.35 to each:
    a. 43.23 b. 80.214 c. 48,903

11. Add 538 + 7.6 + 7.6 + 7.6 + 7.6 + 7.6 + 7.6 + 7.6

12. Subtract 87 from each:
    a. 856 b. 943.74 c. 53,456

13. Subtract: 539 − 78 − 78 − 78 − 78 − 78 − 78 − 78 − 78

14. Multiply each by 6.87:
    a. 685 b. 4,152 c. 79.36

15. Multiply 65 × 4 × 4 × 4 × 4 × 4 × 4 × 4 × 4 × 4

16. Divide each by 5.8:
    a. 2,392.5 b. 4,152 c. 767.05

17. Divide: 10,633,958 ÷ 23 ÷ 23 ÷ 23 ÷ 23

1. _____
2. _____
3. _____
4. _____
5. _____
6. _____
7. _____
8. _____
9. _____
10a. _____
10b. _____
10c. _____
11. _____
12a. _____
12b. _____
12c. _____
13. _____
14a. _____
14b. _____
14c. _____
15. _____
16a. _____
16b. _____
16c. _____
17. _____

TEST A3

Name _____ Total Score _____

Date _____

Use an 8-digit calculator to solve these problems.

For each of these, show the decimal equivalent that should be entered. (2 each)

1. $\frac{1}{6} =$
2. $\frac{1}{5} =$
3. $\frac{1}{7} =$
4. $\frac{3}{8} =$
5. $\frac{5}{6} =$
6. $\frac{1}{12} =$

Solve these problems. Express your answer as a decimal (2 each)

7. Add: $29\frac{1}{8} + 436\frac{1}{2} + 159\frac{1}{4} =$
8. Subtract: $73\frac{3}{8} - 45\frac{3}{4} =$
9. Multiply: $36\frac{5}{8} \times 24\frac{1}{6} =$
10. Divide: $98\frac{5}{16} \div 6\frac{7}{8} =$

Use the constant to add: (3 each)

11. 375 to:
    a. 8,360   b. 964   c. 534.92

12. 5,045 to:
    a. 510   b. 52.141   c. 50,993

13. Add: 450 + 6.8 + 6.8 + 6.8 + 6.8 + 6.8 + 6.8

Use the constant to subtract: (3 each)

14. 78 from:
    a. 134   b. 14,120   c. 712.77

15. 49.8 from:
    a. 60   b. 802.07   c. 84,472

16. Subtract: 829 − 42 − 42 − 42 − 42 − 42 − 42 − 42

Use the constant to multiply each number: (3 each)

17. by 56:
    a. 952   b. 52,786   c. 62.25

18. by 0.09:
    a. 163   b. 0.895   c. 5.877

19. Multiply: 3.7 × 1.5 × 1.5 × 1.5 × 1.5 × 1.5 × 1.5

Use the constant to divide each number: (3 each)

20. by 71.6:
    a. 2,076.4   b. 590.7   c. 32.358

21. by 14:
    a. 63,386.05   b. 6,804   c. 95.41

22. Divide: 45.5625 ÷ 1.5 ÷ 1.5 ÷ 1.5 ÷ 1.5 ÷ 1.5

© Copyright South-Western Publishing Co.

LESSON A4 — Calculator Cutoff, Overflow, and Percent

Cutoff. An 8-digit calculator shows only eight digits in display. For longer answers, the extra digits on the right are *cut off*. The calculator shows the eight digits that are most significant.

Example: 9.8765 × 4.3201 = 42.66746765.
The 6 and 5 are not in display.

Overflow. When the product of numbers being multiplied is too large to fit the display, an *overflow* may be indicated by a flashing display or an E before or after the product. The E means *exponential* notation.

Example:
56,789 × 4,321 = E2.4538526 or 2.4538526E

When overflow occurs on an eight-digit calculator, the decimal point is understood to fall *eight* places to the right. On a 10-digit calculator, ten places to the right. An 8-digit display of E2.4538526 for a product shows that 245,385,260 is as precise as the calculator can indicate.

Percent. If the calculator has a % key, pressing it will place the decimal point correctly. If there is no % key, the operator must remember to move the decimal point two places to the left when removing the % sign and two places to the right when appending the % sign.

Example: 40 × 7.5 % gives 3 *or*
40 × 0.075 = gives 3

EXERCISE A4

Use an 8-digit calculator to compute each problem. Where applicable, round to the nearest cent. (8 each)

1. 86.7142 × 12.803
2. 69,640 × 5,752
3. 32,219 × 9,123
4. 173,681 × 81,014
5. 4,620,000 ÷ 0.00084
6. 71.0147 × 15.2434
7. 687,250 × 498,203
8. 8,720,000 ÷ 0.00075
9. 32.5185 × 75.3494
10. 47,647 × 26,362
11. 353 × 8%
12. 31.5 × 26%
13. 560 × $87\frac{1}{2}$%
14. 512 × 125%
15. 181 × $3\frac{1}{2}$%
16. 630 × $\frac{1}{2}$%
17. $56\frac{1}{4}$% of $88
18. 400% of $25.58
19. $93\frac{3}{4}$% of $480
20. $\frac{7}{8}$% of $501.20
21. Find 7% of
    a. $28
    b. $83.60
    c. 80¢
22. How much is $12\frac{1}{2}$% more than $832?
23. How much is 9% less than $540?

TEST A4

Name _____ Total Score _____

Date _____

Use an 8-digit calculator. Where applicable, round to the nearest cent. (4 each)

1. 84.0329 × 12.3546

2. 765,370 × 487,305

3. 56,789 × 4,321

4. 874,018 × 468,141

5. 9,650,000 ÷ 0.00085

6. 52.2185 × 35.7494

7. 137,816 × 8,132

8. 5,139,000 ÷ 0.000625

9. 59.8545 × 73.7632

10. 69,460 × 6,257

11. 174 × 9%

12. 647 × 374%

13. 5 × 7%

14. 48 × 30%

15. 188 × $37\frac{1}{2}$%

16. 6,730 × $\frac{1}{4}$%

17. $6\frac{1}{2}$% of $98

18. 432% of $850

19. 0.8% of $840

20. $\frac{5}{8}$% of $710.20

21. Find 8% of

   a. $84

   b. $136.80

   c. 90¢

22. How much is $16\frac{2}{3}$% more than $732?

23. How much is 15% less than $860?

1. _____
2. _____
3. _____
4. _____
5. _____
6. _____
7. _____
8. _____
9. _____
10. _____
11. _____
12. _____
13. _____
14. _____
15. _____
16. _____
17. _____
18. _____
19. _____
20. _____
21a. _____
21b. _____
21c. _____
22. _____
23. _____

A Cumulative Review

Name _____

Use an 8-digit calculator to solve these problems. Write each answer in the appropriate space on the right.

1. 93,452 + 10,854 − 43,059 − 66,159 =

2. 54,725 − 97,942 + 84,235 − 6,853 − 92,452 =

3. −645 + 926 + (−660) + 629 =

4. −546 + 899 + (−930) + (−914) + 837 =

5. 868 × 53 × 45 =

6. 7,848 × 84 × 86 =

7. 9,566,250 ÷ 76,530 =

8. $\dfrac{10,030,236}{3,658} =$

9. 441,558 × 38 ÷ 481 =

10. $\dfrac{57,021 \times 354}{687} =$

Write the on-line form for each of these.

11. $\dfrac{1,288 \times 5 \times 6}{8 \times 7} =$

12. $\dfrac{(34 + 8) \times 4}{14} =$

Solve each of these equations.

13. $n = 57 + 31 \times 45 \div 15$

14. $n = \dfrac{336 \div 4 + 51 \times 35}{15 + 6}$

Use the memory register of your calculator to solve these problems.

15. $n = (54 \times 71) + (67 \times 53)$

16. $x = (47 \times 82) − (58 \times 45)$

17. $y = (207 \times 92) \div (69 \times 23)$

18. $z = (675 \times 162) \div (135 \times 54)$

19. Add: $78\dfrac{5}{8} + 325\dfrac{1}{2} + 279\dfrac{1}{4} =$

20. Subtract: $64\dfrac{2}{3} − 42\dfrac{1}{6} =$

21. Multiply: $72\dfrac{1}{12} \times 34\dfrac{3}{8} =$

22. Divide: $112\dfrac{7}{8} \div 8\dfrac{1}{16} =$

1._____
2._____
3._____
4._____
5._____
6._____
7._____
8._____
9._____
10._____
11._____
12._____
13._____
14._____
15._____
16._____
17._____
18._____
19._____
20._____
21._____
22._____

A Cumulative Review

Add 378.65 to each number:

23. 593

24. 73.084

25. 4.2789

26. Add: 847 + 65.92 + 65.92 + 65.92 + 65.92 + 65.92 + 65.92

Subtract 85.49 from each number:

27. 4,312

28. 342.64

29. 78.3

30. Subtract: 669.79 − 72.8 − 72.8 − 72.8 − 72.8 − 72.8 − 72.8 − 72.8

Multiply each number by 4.68:

31. 967

32. 78.78

33. 8.0605

34. Multiply: 46 × 1.4 × 1.4 × 1.4 × 1.4 × 1.4

Divide each number by 17:

35. 72,474.995

36. 95.2

37. 0.0119

38. Divide: 195.3125 ÷ 2.5 ÷ 2.5 ÷ 2.5 ÷ 2.5

39. Multiply: 162,507 × 35,473

40. Divide: $\dfrac{97{,}255{,}000}{0.0005}$

23. _____

24. _____

25. _____

26. _____

27. _____

28. _____

29. _____

30. _____

31. _____

32. _____

33. _____

34. _____

35. _____

36. _____

37. _____

38. _____

39. _____

40. _____